Foreword by ELLIOTT MASIE, Author of BIG LEARNING DATA

LEARNING IN THE AGE OF IMMEDIACY

5 Factors for How We Connect, Communicate, and Get Work Done

Brandon Carson

20 19 18 17 1 2 3 4 5

ATD Press is an internationally renowned source of insightful and practical information on talent development, workplace learning, and professional development.

ATD Press
1640 King Street
Alexandria, VA 22314 USA

Ordering information: Books published by ATD Press can be purchased by visiting ATD's website at www.td.org/books or by calling 800.628.2783 or 703.683.8100.

Library of Congress Control Number: 2017937927

ISBN-10: 1-56286-769-5
ISBN-13: 978-1-56286-769-0
e-ISBN: 978-1-56286-982-3

ATD Press Editorial Staff
Director: Kristine Luecker
Manager: Christian Green
Associate Director, Communities of Practice: Justin Brusino
Developmental Editor: Kathryn Stafford
Senior Associate Editor: Caroline Coppel
Text Design: Iris Sanchez
Cover Design: Studio Carnelian
Printed by Data Reproductions Corporation, Auburn Hills, MI

For Hannah

Contents

Foreword

I WAS RECENTLY PREPARING TO GIVE A speech at an oil company in Saudi Arabia and playing with my new Snap Spectacles. Dozens of learning and business leaders flocked forward to touch, try, and explore the new wearable glasses that playfully capture 10-second videos and automatically share them with friends on Snapchat. We were experimenting with tomorrow's technology, today, in a social media context—changing the nature of innovation. An important realization for me and this audience was that future learning technology may no longer require a major system, significant expenditure, or expensive expertise.

Today, modern learning organizations are on the cusp of radical evolution and revolution. In business terms, technology has brought massive transformation in how we get things done. It's no longer possible for learning leaders to be stationary or rely on traditional methods to support their workforce.

This book is a primer on what Brandon Carson refers to as the five factors every learning leader needs to know as they evolve their organizations to move at the speed of business. *Learning in the Age of Immediacy* is an excellent guide for helping you begin to understand emerging trends, platforms, ecosystems, and the knowledge-building innovation that is requiring us to reimagine our learning strategies. The book explains in real terms how to best wrap your head around emerging technologies that are driving a wholesale transformation of how people communicate and learn. Brandon offers real-world examples, case studies, and practical approaches to help you decide whether you should integrate these factors into your strategies. This book has reignited my belief that technology will always play a significant role in any learning strategy, and it's exciting to learn more about how these technologies, alone and combined, are really about learning. I now understand why the modern learning organization should never stop learning either.

Some important concepts explored in this book include:

- Automation in the workplace will radically transform how we work. We are just now entering an era of intelligent, connected devices that will forever alter the meaning of "work." It's imperative for learning organizations to know more now about the impact of automation. Here, you'll learn about intelligent machines, immersive learning experiences, and how machines and humans work and learn together.
- The cloud has revolutionized how we communicate and collaborate. As Brandon states, "we're still in the early stages of our move to the cloud, but it's already obvious that [this factor] is a game changer for learning delivery." Every learning leader needs to know and embrace cloud technology.
- Of the five factors Brandon discusses, mobile is the one with which we are all probably the most familiar. He discusses how we need to be thinking mobile-first now, shows us how mobile has consumed us all, and how it's not just about a device in a hand. Mobile is how everyone gets their work done. You'll be surprised by some of the observations about how mobile has accelerated our ability to build more relevant and meaningful learning experiences.
- Learning leaders also must foster more data-driven functions. Big data and learning analytics seem daunting at first, but you'll learn how establishing a data strategy is the first step to understanding your impact on performance and business imperatives. Data are everywhere, and you now can extract meaning and guidance from them.

- Finally, Brandon dives in to the nascent but quickly growing Internet of Everything. He breaks down the meaning and explains the key points for learning organizations to focus on when it comes to harnessing our new world of "always-on and always-connected."

Let me impress upon you that this book is not about hype and buzzwords. Brandon is more than just a fanatic early adopter. The five factors discussed may be emerging technology, or technology at an early stage, but all are already changing business models and enabling businesses to do more, faster. Each factor requires you to rethink how you design and deliver learning and, more important, helps you learn how to extend, widen, enrich, and deepen your contribution to the workplace you support.

After I wrapped up my speech in Saudi Arabia, the folks in the room circled around me, waiting to try on the new wearables. The discussion quickly evolved from excitement around a new gadget to how to incorporate them into their offerings to provide new learning opportunities. Now is the time to embrace the current emerging shifts that are becoming our future. Welcome to the Age of Immediacy.

Elliott Masie
May 2017

Preface

"I don't need a hard disk in my computer if I can get to the server faster. . . . Carrying around these nonconnected devices is byzantine by comparison."
—Steve Jobs, 1997

I HAVE BEEN USING DIGITAL TECHNOLOGY SINCE I was 15, when I was introduced to it by my first computer: a Radio Shack TRS-80. I learned how to program in BASIC, which started me on a technological journey that continues today. Until recently, I always considered technology separate from my day-to-day life, not interwoven into every aspect of it. I think we can all agree that has changed. Over the last two decades, technology has transformed our lives and accelerated our state of connectedness. It's now imaginable for us to accept the idea that every person, place, and thing will eventually be connected, creating a vast network of everything.

I've spent the last 25 years in the learning industry as an instructional designer, strategist, and leader focused on how to improve the performance of the workforces I support. I strive to leverage technology resources when appropriate to build meaningful learning experiences that make people's work lives better. I've been fortunate to experience the wholesale transformation of the learning industry over the years, and watch and participate in how almost everything we do has evolved. Along the way, some of what we valued most has been questioned, such as the efficacy and ROI of classroom training. Some delivery methods, such as mobile learning, caused us to challenge our thinking of effective learning methods, but has now evolved into an important part of the learning toolbox.

The pace of change in the learning industry has progressed remarkably, and in recent years has begun to play a more central role in strategic planning at the C-suite level. In fact, many CEOs are relying more and more on their chief learning officers (CLOs) to help identify and measure workforce capabilities and align those to the needs of their business.

In recent years, technology has been the driving force behind much of the business transformation now occurring, creating completely new industries, and in some instances, remaking existing industries entirely. When the story of how technology disrupted the workplace in this era is written, it may be viewed from three distinct waves (based on Toffler 1980):

- **Wave 1: The Internet Revolution (1989-2000)**. The rise of the Internet brings rapid technological change. The first focus is on developing the systems to connect people and standards for commerce and massive information dissemination.
- **Wave 2: The Information Revolution (2000-2025)**. Where we are right now. Information becomes accessible in near real time, changing our content consumption patterns and behavior, and transforming education and how we acquire knowledge.
- **Wave 3: The Social Revolution (2025-)**. There are no unconnected people on earth. We are less defined by physical geography, and more by our online activity.

As a learning leader, the forces of change affecting business require you to be more entrepreneurial, to navigate unchartered areas, and to make decisions in a space where skepticism may be high. For some, the barriers to successful integration of emerging technology with learning strategies may seem far greater than anything ever experienced. In the face of constant change, the modern learning organization has a responsibility to guide the business and workforce forward, providing counsel to executive leadership teams on workforce capability and evidence that the workforce can execute on the business imperatives.

The learning industry is at a fork in the road, and it's important to view how your learning organization functions from two viewpoints: a startup entrepreneurship and an enterprise mindset. A nimble, responsive learning organization builds capabilities that the business identifies and leverages for competitive advantage. The growing demand for knowledge sharing, collaboration, and problem solving requires the ability to focus on the interdependency of all the business functions. It's important to perceive your learning business as one that needs to attract investors; is it positioned well to lure skeptical venture capitalists? You must be able to tell the story of where you're taking your learning organization and how you're providing value, and show the evidence supporting that narrative.

I hope the information in this book helps shape your thinking around some of the fundamental aspects of reimagining your learning business and more effectively position it to take full advantage of the forces of technological change. The stakes have never been higher—the choices you make now will either extend your learning organization's ability to focus on real outcomes and measurable results, or you will fail to deliver actionable results, requiring those you support to seek solutions elsewhere.

Acknowledgments

THIS BOOK IS THE RESULT OF A multiyear labor of love, beginning with observing how learning organizations navigate the complexity of modern-day business. I constantly compiled notes and ideas and delivered presentations at conferences and to other learning leaders on emerging technology and its effect on the learning function. Those around me suggested that the output of those notes, ideas, and presentations might make an interesting book, which ultimately led to this publication. I would like to thank these folks for supporting and advising me during the writing: Michelle Lentz for helping me get the words on paper in a way that makes the most sense; Penny Carson for making sure I stayed on target and consistently challenging me on my explanations and perspectives; Elliott Masie for his advice and support; and Richard Barr for being a nonindustry sounding board. I'd also like to thank Justin Brusino and Bridget Dunn from ATD for having the vision to take this project forward, and Christian Green and Caroline Coppel for their editing prowess.

Introduction: Welcome to the Age of Immediacy

WHICH WAS A MORE IMPORTANT INNOVATION: INDOOR plumbing, jet air travel, or the Internet? Each of these innovations ushered in new eras of productivity and human capability. Similar to air travel, but at an even more accelerated pace, the Internet has developed astoundingly fast, affecting every aspect of our lives, including the technological, political, social, and educational landscape. It has affected how we communicate, collaborate, congregate, and most important, learn. Over time, the Internet may stand as the most significant human achievement. How it evolves is hard to predict, but in less than 40 years it has completely altered how we experience our world. The Internet revolution was driven by the rise of easy-to-use, connected devices and near-ubiquitous access to networks. How we acquire and share knowledge has been fundamentally changed, requiring us to reevaluate the very construct of our training systems and frameworks. We are speeding head-on into the Age of Immediacy, where we expect information to be available when we want it, regardless of where we are, and in an easily consumable format.

You probably chose this book because of your role in learning and your need to address the turbulence brought on by the Age of Immediacy, including decisions regarding technology, strategic vision, tactical execution, and workforce development. It may be your job to support the business as it moves away from business as usual, which means you have been placed in a position requiring you to rethink every aspect of your learning business.

This book is not about glorifying digital technology or imploring you to implement each of the five technology factors discussed. Instead, it's about helping you recognize how, in such a short time, digital technology has fundamentally affected almost every aspect of how business is conducted and has altered how we should be designing, developing, and delivering learning experiences. Much of what has happened over the past few years is a precursor to what will happen over the next 30. We are still in the formative stages of some of these factors, but it is certain we will be interacting with these technologies in all we do as everything on earth becomes connected.

From workplace automation, including robots, chatbots, artificial intelligence, and virtual and augmented reality, to the ascendance of the cloud and mobile technology, to the influence of big data and analytics and the Internet of Everything, these factors have begun to disrupt both business and learning organizations. Although I discuss each separately, they overlap; some are dependent on others, and at times it's a challenge to discuss one without another. However, it's important to look at each factor separately so you can gain more insight into the uniqueness of each.

Workplace Automation

Workplace automation is a topic that both entices and frightens people. The idea that computers, robots, and algorithms could make us obsolete and irrelevant is enough to make us wonder about the future of humanity itself. Automating job tasks is not a new paradigm. Our society has worried about how technology disrupts our livelihoods since the turn of the 20th century. This era is different. With advances in machine learning now moving at an accelerated pace, it won't be long until software can do much more than just react to input. Software will anticipate our needs, complete our tasks, and connect us to faster, simpler ways of working. For many tasks, software and machines will replace humans, which begs the question: What will the workforce you support look like in the next several years? Are the skills and capabilities you have in your organization the same ones you will need to future-proof your team? Just look at Blockbuster, Nokia, or Blackberry, each of which had opportunities to transform their businesses to prepare for or anticipate changes brought on by technology. But they were slow to respond to the technological disruption that affected their businesses. Where are they now? Workplace automation is changing how you need to respond from a learning perspective, but it's also affecting the talent you need in areas such as collaboration, creativity, strategic and analytical thinking, and work intermediation (ability to integrate digitization into almost any work method). Your learning team cannot respond proactively if they do not have the technical acumen to provide the right learning solutions based on the rapid rise of automation in the workplace.

The Cloud

When asked about the cloud in 2008, Oracle founder Larry Ellison exclaimed, "What is it? It's complete gibberish. It's insane. When is this idiocy going to stop?" (Farber 2008). At the time, Ellison believed the cloud was no more than a buzzword. But as it has evolved, cloud computing has come to represent the perfect aggregation of technology and services driving our insatiable quest for real-time, contextual knowledge. As we begin to evolve from the ownership economy (owning or storing our data on our own hard drives) to the streaming economy (accessing data stored elsewhere on demand), we are seeing explosive growth in content decentralization. We can now access data from any device at any time from remote servers while we are connected. The cloud has become a utility that has changed the nature of how we interact with one another, transformed multidevice computing, and exponentially increased our ability to learn, share, and acquire new skills and knowledge in almost any context.

Mobile

Mobile technology is arguably the factor that has had the biggest influence on the workplace thus far. With mobile devices, our behavior has dramatically shifted. Now we're "always on" and our experiences are more direct and hyper-individualized. Information is only a tap away. This is why we have become emotionally attached to our devices—we touch and rely so much on them, so we form a deeper relationship. We expect our devices to assist us through almost every aspect of our lives: how we communicate to one another, research and purchase products and services, conduct work, schedule our day, and yes, learn. We expect our devices to entertain, but also shepherd us through our tasks. They are always with us and keep us connected. The mass adoption and major behavioral change brought on by smartphones is the primary reason you need to rethink everything you're doing when it comes to your learning strategy.

Big Data and Analytics

For training, the impact of big data and analytics is twofold. First, it provides more complete information about what your learners are doing, where they're excelling, where they're struggling, and how they're actually using your content. Second, big data can help predict learner behavior. You want a workforce that can perform to the capabilities the business needs to execute on its goals. Your challenge is knowing whether the workforce is capable of achieving those business needs. Every time someone does something with their connected device, you can gather data.

The emerging practice of data science and behavior prediction is leveraging these data to find trends and analyze those trends to predict action. You want to have the capability to know what action is about to happen so you can step in and change any action that is not desired. You also want to interpret both formative and summative data to inform your training strategy.

The Internet of Everything

The growing Internet of Everything is emerging as the grand connector that enables all our devices to communicate with one another and us. The promise of a more connected ecosystem is to truly realize the idea of contextual learning. Imagine a day when workplace knowledge is less about pushing information through one centralized system, such as your learning management system (LMS), but instead focusing on decentralizing content. Learning content becomes more of a feed delivered through multiple channels at any moment of need—think smart displays in the work environment, signage, product packaging, mobile devices, wearables, sensors, and so on. Everything will be capable of connecting to the network and assisting. Think of how sensors communicating to other sensors will craft a customized experience specific to the context of the learner.

***

The primary goal of this book is to help establish a frame of reference for these factors so you can decide whether and how to integrate them into your learning strategies in both the short term and the long term. My hope is that the information presented will aid you in making strategic choices based on what works best for your organization, its values, culture, and processes.

Why This Matters Now

While technology alone doesn't improve training, it's become our permanent partner in how we design and deliver learning experiences and evaluate their effectiveness. Increasingly, the workforce we support must learn new skills, innovate and create quicker, and boost performance while absorbing rapid change in how they communicate and collaborate to get their work done. Unlike previous technological disruptions, such as radio and television, the Internet has always been about handing the power of communication to the masses; no invention has spread as quickly and transformed our society as much in such a short time.

The five factors discussed in this book are or will affect how training is designed, delivered, and evaluated. These factors, sometimes referred to as "edge technologies," consist of emerging technology products, services, and trends that are predicted to have a significant impact not only on business but also on workers and their performance. It's important to be aware of these factors so you can appropriately determine their place, if any, in your organization. These five factors will cause the biggest transformation in how we connect, communicate, and get work done. Many other technology factors disrupt business, but as an aggregate, these five will have the most influence over the next several years. The irony is that each factor offers both interesting opportunities and daunting challenges for typical learning organizations.

One of the biggest hurdles to overcome is the realization that these factors aren't really meant to help you do things the way you always have; they're not just about making it easier to implement your existing strategy. These factors demand a new way of thinking about how you should approach the business of learning in its entirety. Many learning leaders strive to find a way to do more of the same with increased efficiency, productivity, and cost effectiveness. John Hagel (2014), co-chair of Deloitte's Center for the Edge, a management research consulting firm, has said that when it comes to digital strategies, executives need to "fundamentally step back and rethink what business they're in." Now, more than ever, it's time to rethink talent needs across learning teams. Jobs are evolving, technology is embedded in almost all work tasks, and the right talent will be scarce as skill sets evolve. Your team needs the capabilities to proactively deliver relevant training across the spectrum of these changes.

Our newfound ability to instantly share and receive information with anyone, anywhere means we are now in a world of not only real-time information, but also real-time learning. How does this affect your learning organization? Never again will anyone expect to *wait* to learn something. Can you support your audiences with up-to-date, meaningful content? Are you able to make that content easily discoverable, on demand? Do you have the skill sets and resources on your learning team to design and deliver device-optimized content to your audiences?

Through conversations with industry experts and explorations into emerging technologies and services, I hope to help you gain a deeper understanding of the importance of knowing what's coming next, to make more informed decisions about which factors may be right for your organization. For your learning team, the next few years are the beginning of everything they do changing for the better.

1

Workplace Automation

"I do not fear computers. I fear the lack of them."
—Isaac Asimov

TRACY PHILLIPS IS A SENIOR BUSINESS ANALYST at a major accounting firm. She leads a team responsible for monitoring thousands of the firm's accounts. They scan activity across the accounts, proactively flagging any with errors or activity that may cause an audit or investigation. Her team consists of only two other analysts.

Tracy's team can process more than 2,000 accounts on a continuous basis because of workplace automation. Before the account monitoring system was automated, a large number of analysts were required to observe the same number of accounts. By implementing a system to automate almost 80 percent of the tasks, the firm was able to reduce the number of humans required to do the job. This has also resulted in increased accuracy, meaning fewer audits for her clients.

***

Over the last few decades, digital technology has changed almost every aspect of our work and how we do it. From desktop computing, the Internet, and connected devices, to intelligent machines such as driverless cars and autonomous robots, many experts are predicting that much of the work currently being performed by humans will be replaced by automation. Workplace automation–

the use of intelligent systems to replace or reduce human intervention—has the potential to affect almost 50 percent of all jobs in the United States (Smith 2016). However, technological advancement is not the only reason many jobs are being automated. The transition to workplace automation is a convergence of three massive shifts occurring simultaneously:

1. **Major structural readjustment to business brought on by globalization and technology**. Many of the processes in manufacturing, distribution, and the supply chain are being readjusted to accommodate the global market. Robotics and analytics stand out as offering near-term gains in productivity and efficiency. Currently, these factors are leveraged largely in manufacturing, allowing businesses to expand their target markets, but robotics and analytics will eventually span many business operations. Technology is also a crucial enabler in market expansion. By moving to ubiquitous connectivity, businesses are able to concentrate economic activity in a time-sensitive and highly customized fashion to deliver products and services in record time. Over the next several years, technology will displace specific workers while creating highly optimized and more efficient work systems. Industry will focus on finding the right balance between the human-technology dynamic by identifying the tasks that are best automated and those that require humans.
2. **Changes in the high-tech post-industrial economy**. A more integrated global economy affects almost every business system, and reconfigured work processes influence tried and true ways of getting work done. Many of the systems and tools once used to perform day-to-day work tasks have already begun their descent into obsolescence. The desktop computer, a relatively recent invention, is a prime example of a prominent work device rapidly becoming outdated as it's overtaken by portable connected devices. The last decade has seen the beginning of untethering from wired, dedicated connections to always-on and persistent wireless access.

Businesses will begin to evaluate their technology systems from back to front, seeking efficiency enhancements and optimization of internal processes. The emergence of the platform as a conduit to increased collaboration and communication on a global scale will make it possible to almost instantly expand markets and test opportunities without significant risk exposure.

3. **Evolution of the workforce**. The largest worker demographic is getting older and beginning to retire, creating a labor shortage. In 2016, almost a third of the U.S. workforce reached age 50 or older—a group that is expected to grow to 115 million by 2020 (Harrington and Heidkamp 2013). The coming labor shortage paints a tough picture for business. Maintaining skilled workers at all levels will become a challenge, especially for manufacturing, healthcare, and service-oriented businesses. However, technology will spur innovative business practices that alleviate the labor shortage, and also potentially attract older workers back to work. Providing options in how they can work should catalyze business, restore balance and sustainability, and open new types of jobs.

It's increasingly difficult to predict the skills needed in the workforce to keep up with the pace of change. Increased automation of work currently being done by humans will have a profound impact on learning and development for companies that leverage digital technology to automate their workplaces. Currently, workplace automation is mostly affecting blue-collar workers. As these workers are displaced by automation, you won't need to train them on how to perform the tasks to which they are accustomed. You will, however, need to offer training on how to program, monitor, and service the devices taking over the tasks. Consider how ATMs are replacing bank tellers, and how automated check-out systems are reducing the need for cashiers. These automated systems and processes are creating new jobs in engineering, customer service, and management. If one cashier has to monitor, service, and maintain five automated point-of-sale systems, he will need advanced training on troubleshooting, accuracy, and precision, as well as the ability to help customers interact with the technology. Management will need training on how to better coordinate worker activities, understand the technical aspects of the systems, and help redirect displaced workers to other functional areas of the business.

Workplace automation is beginning to affect white-collar jobs as well. Travel agents are disrupted by travel-planning websites; print journalists are being displaced by the Internet, and in some industries, the need for salespeople is dwindling because of online personalized shopping options. You may need to provide more technical and leadership training as these workers transition to higher levels of responsibility. Managers will need advanced skills on how to lead mixed teams—machines and humans working side by side. Workers will need deeper analytical skills and the capability to both gather and interpret data generated by the automated systems.

Even the basic flow of information has expanded to enable real-time communication and collaboration. And this has all happened in just the last few decades. It's difficult at times to comprehend the amount of change that's occurred in the workplace since the introduction of the iPhone just a little more than a decade ago. Consider that it took the telephone almost 80 years to become a ubiquitous device. One of our biggest challenges will be unlearning old practices and processes as the workplace becomes more automated and driven by technology.

Let's take a deeper dive into three emerging technologies currently redefining tomorrow's workplace: robots, artificial intelligence, and virtual and augmented reality.

Rise of the Robots

It's a sunny afternoon, and Jenny wants to escape the confines of her cubicle, eat lunch outside, and take a breather from her busy morning. Today she decides to try a different approach to lunch in the park. She finds a vacant bench, sits down, and taps the Eat Now app on her phone. Jenny orders a fresh, hand-crafted lunch to be delivered directly to her bench in less than 20 minutes.

Most of us have been ordering food on demand, such as pizza and Chinese, for years. Send in the order using the phone, the web, or an app, and someone brings it to you in minutes. The difference here is a human won't be delivering Jenny's lunch. Instead, it's coming to her by a six-wheeled robot, or ground drone. The robot is covered with cameras and sensors that it uses to navigate (at four miles per hour) from the restaurant to Jenny's park bench. Once it arrives, Jenny enters a code, and the robot's lid opens so she can retrieve her food. As she enjoys her lunch, Jenny takes a moment to marvel at what just occurred: having not prepared anything for lunch that morning, she impulsively decided to dine in the park. With four simple taps on her phone, she sent a lunch order to a local eatery where a fresh meal was quickly prepared, wrapped, and loaded into a waiting robot, which then self-navigated the streets to deliver the food to her exact location. This interaction represents a perfect example of how humans and robots can work together to transform traditional business services and processes.

Automation is best used where tasks are repeatable and predictable. In this example, the unmanned ground drone was able to move along city streets using cameras and sensors, but it was still monitored by humans in the company's control office. Computers don't yet have advanced critical thinking skills. The ground drone can easily get from point A to point B, but by itself it's not very flexible at making decisions. Its mission is not to "take Jenny her lunch"; its mission is to move from one coordinate to another without interruption. Creativity and flexibility (such as if the drone is interrupted by a stranger or object or chased by a dog) are critical skills that automation won't be able to replace for quite some time (Hoffman 2016). These tasks are left squarely for humans to perform, although there are rapid breakthroughs happening in this area.

Digital technologies are accelerating automation in the workplace and taking over the tasks at which humans toil away, with speed and accuracy far above what humans are capable of doing. These automated systems are helping companies get products and services to market at unheard of speeds.

The areas in which robots are most often found right now are manufacturing, construction, healthcare, and the supply chain. In fact, supply chains are becoming one of the most disrupted areas of the business, as robots take on more of the laborious work.

We've all heard the term *robot*, of course, but you may find it interesting to know where the term originated. Almost 100 years ago, in 1920, *robot* was used for the first time in a play written by Czech writer Karel Capek (Wood 2015). The play, *Rossum's Universal Robots*, described artificial persons and had both male and female robots. He based the term *robot* on the Czech word for forced labor. The use of the word progressed through various dystopian novels, plays, and movies, eventually becoming a science fiction staple.

Now, when you think about robots in the workplace, more than likely you envision a manufacturing plant where they are hard at work assembling products. The odds are high that your life has been affected in some way by a robot: the clothes you wear, the food you eat, or the car you drive were most likely worked on or assembled by robots.

As robotic technology evolves, different types of robots are integrated into the workplace. The most common are industrial and service robots, but there are multiple variations of other types emerging, including small-scale robots, flying micro-robots, and even soft robots.

Industrial Robots

The most common type of workplace robot is the industrial robot. Industrial robots are usually segregated from humans because they can be large, heavy machines with significant load-bearing capability and potentially pose a safety risk. The first industrial robot was put into use in a General Motors plant in 1961 in Trenton, New Jersey, to handle die cast metal and to stack heavy materials. The use of industrial robots has become standard for manufacturing products such as cars, electronics, and large machinery. When compared with humans, industrial robots return the highest productivity, reliability, quality, and cost effectiveness.

Service Robots

In the 1990s, service robots emerged. Service robots are meant to have direct interaction with humans—the robot that served Jenny her lunch is an example of a service robot. Service robots are used in industries such as hospitals, warehouses, and hotels. They are meant to perform mundane tasks such as stocking hospital supplies, fetching retail orders in a warehouse, and delivering room service to hotel guests, yet still maintaining the human touch because they work alongside and interact with people. Service robot mishaps are difficult to predict; because they work in nonstructured environments, can use vision systems to move about, and are able to grasp and hold objects, there is a risk that they could hurt humans or cause damage. They may fail to stop or accelerate at a high speed, or a robot's arm might engage in an abrupt, unexpected motion hitting a human or causing damage to the physical environment. However, the reality is that humans who operate robots are still more prone to error than the robots themselves (Vasic and Billard 2013).

Other Robot Types

Small-scale robots, even ones the size of a fly, are used to manufacture high-precision products like computer chips with much more accuracy than that of humans. Leveraging these robots to manufacture small systems lessens the need for humans to toil at surgical-like skills, where we are imprecise at best. Small-scale robots can also work cooperatively to build large-scale objects. Small-scale robots can be woven into textiles, providing assistance to enhance performance (such as athletic performance) or to aid those with physical impairments.

Flying micro-robots can be made quickly and in bulk. They are used in hazardous environments and for search and rescue operations. These micro-robots can also be used in medical surgery where they can fly or crawl into cells.

Soft robots are less rigid and built with soft materials, preventing them from hurting humans. They can be integrated into almost any environment because they pose fewer safety risks. They are more likely to be used as assistive devices and in rehabilitation for physical impairment.

The Human-Robot Alliance

Robots are predicted to become a critical part of the workforce across many industries, including transportation and logistics, customer service, home maintenance, and as stated earlier, healthcare. Japan and the United States have been leveraging the use of robots more than other areas of the world. In Japan, the use of industrial robots is astounding. It's estimated that by 2025, robots will shave 25 percent off Japan's factory labor costs (Yoshiaki 2015).

Often, our first reaction to robots is to worry about the displaced humans and the adverse effect on employment. However, what's really happening is robots and humans are beginning to work side by side. This trend is most obvious in distribution centers. The use of robots to process and fulfill orders is directly affecting business productivity, while freeing up humans to innovate processes and focus more on business-critical skills such as customer service.

This human-robot alliance is beginning to alter the landscape of the workforce: robots are better at performing repeatable, rote tasks while humans are better at performing customized, more creative, and innovative work. Human productivity is best augmented, not replaced, by robots to help increase productivity and flexibility. Humans can see, select, and grasp with more agility; robots can handle, lift, and move heavy items repeatedly without tiring or getting bored. Together, robots and humans do more with increased speed and accuracy.

The trend toward collaborative human-robot work sharing represents the first wave of automating tasks, which will result in massive business productivity gains. For example, robots working alongside humans in warehouses are increasing warehouse productivity by enabling faster shipping, which will eventually allow retailers to deliver products within hours and minutes versus days. Other robots are helping increase customer service and satisfaction by assisting customers with locating products. In many situations, robots that speak multiple languages and use facial recognition can get a person to the product they're looking for much faster than a human can.

What does the rise of robots in the workplace mean for learning? Will we have a workforce of humans teaching robots how to do their jobs? Or the other way around—robots training humans to work? The idea of robots training humans seems strange, but it is foreseeable. As multipurpose robots become more integrated in the workplace, humans will naturally interact with them more, issue commands, listen, observe, and at times, learn from them. Robots are also capable of self-teaching. Software is the key driver enabling them to teach themselves what they need to do. Deep learning, often referred to as machine learning, provides mechanisms for machines to teach themselves new skills when exposed to data. Robots use deep learning to acquire skills by collecting data as they work and improve their own performance in real time, which can potentially lead to robots teaching us how to be better at our jobs (Levine, Lillicrap, and Kalakrishnan 2016).

FANUC, the world's largest maker of industrial robots, is applying deep learning as a way to train robots in new tasks, such as learning how to grasp unfamiliar objects (Tobe 2016). When a robot picks up an object, it has a camera attached that captures video of the process. Each time the robot succeeds or fails, software tells it to remember how the object looked, so that next time, the robot can use knowledge of that object to inform how it controls the action. After a number of hours, the robot achieves a higher degree of accuracy equivalent to having had a human program it. This type of learning simplifies and speeds up the programming of robots that perform factory work. When humans no longer have to teach robots, they can be more productive at other tasks robots are not great at performing, such as people management and leading teams, assessing and applying specific expertises to work processes, and engaging in unpredictable work efforts.

Moving away from the world of supply chain, manufacturing, and distribution, you'll find automation in other areas. The increasing complexity of work in general is forcing the industry to look at creative methods to mitigate labor shortages, uncover cost efficiencies, and find more ways to assist workers.

A significant number of tasks for which you currently provide training will not be required, freeing you to focus your resources on other types of training. Automating rote tasks will potentially mitigate the need for producing more and more procedural training. Because your human workforce will be performing less strenuous work, your learning strategy may focus more on innovation, creativity, and how to apply critical thinking to higher level skills that will be key areas of responsibility. From this perspective, the rise of robots in the workplace can be seen more positively than negatively. Just like in the Industrial Age, and even more recently in steel mills, automation replaced many jobs, but humans have an uncanny ability to redirect their efforts and find new things to do.

Ask the Expert: John Hagel on the Future of Work

John Hagel, co-chairman for Deloitte LLP's Center for the Edge, is an expert on how technology is affecting the future of work. Hagel discusses how automation is affecting the workplace, how the forces of technology are reshaping business, and how the idea of work itself is being redefined.

It used to be the label of knowledge worker was applied primarily to high-tech workers. Given how technology has evolved, do you think this label still applies?

First of all, I challenge the label of knowledge worker being applied to just high-tech workers. It's revealing how we think about the evolution of work by singling out a small segment of the workforce and labeling them as knowledge workers, while labeling everyone else as not knowledge workers. For all of us to have more impact, we need to find ways to leverage knowledge and develop it faster regardless of the work we perform. If we're a janitor, factory worker, a scientist, or a researcher, we are all knowledge workers. One reason we've made this distinction is that for centuries work has been structured as an institutional model, based on the idea of scalable efficiency, which is the notion that the highest value from the workforce is efficiency at scale. The best way to get that is to tightly specify all the activities, standardize them, and then tightly integrate them. Huge, multivolume process manuals were created to prescribe how and when the worker should perform tasks. In that environment, what is the role of knowledge? None, just follow the instructions. There are a few privileged people in research or product development that can be knowledge workers and less constrained by the process manual, although the process manual can dominate there as well.

My belief is that the scalable efficiency model is fundamentally challenged by technology. Computers can perform tasks that are highly scripted and tightly structured much more efficiently than humans. We get distracted, we make mistakes, and we get sick. Being an optimist, I do believe technology is a catalyst to rethink work at its most fundamental level, and question whether the scalable efficiency model is what humans should do. I would argue there's a different form of work at which humans excel. The transition to evolving technologies such as artificial intelligence in the workplace will be extremely painful. A lot of people will lose their jobs in the short term. We have institutions that are tightly built on the scalable efficiency model, and they won't rethink work quickly. The transition could potentially be violent as people feel more and more threatened as their jobs disappear.

Do you think technologies such as artificial intelligence (AI) will have significant impact on the workplace?

In the short term, I view artificial intelligence as helping workers alleviate rote, administrative tasks. Machines are already learning on their own. The trajectory with machine learning has been interesting. AI came into being in the 1970s and 1980s with the popular view that it would transform our world. However, it wasn't long until technology challenges appeared, bringing about a period of disillusionment. It's only been in the past few years that we've suddenly seen a resurgence of machine learning capability. Dramatic advances, including the machine outwitting the human chess master, the machine competing and winning at Jeopardy, and so on, are capabilities we never expected to happen so quickly. I do believe machine learning is going to evolve much faster than we expect.

What, in your opinion, should learning leaders be focused on to future-proof their ability to support the workforce of the future?

This is a major focus of my work. Much of our perspective is framed by the long-term forces reshaping the business world. One key consequence is how technology improvement dramatically accelerates the pace of change in the workplace. Training programs have become more and more marginal to learning. By the time you've managed to document processes and procedures, create training materials, and deliver them, the world has already moved forward. I think the real opportunity for training to be relevant is to do less in the training room and provide more support on the job and in the work context. We need to redesign the work environment so it can accelerate learning and performance improvement, not just the physical layout but also the platforms and tools that workers interact with. The entire worker experience needs to be redesigned to accelerate learning in the workplace. Some companies have taken elements of the environment and modified them to increase performance. One big opportunity is around what is generally known as exception handling. Some informal surveys show that 50-70 percent of employee time is consumed by exception handling (addressing things that were not supposed to happen, not part of the policy, and that need to be resolved). Typically, an individual has to spend significant amounts of time seeking resources to help resolve the matter. It can be an inefficient part of work. This is a huge opening for training—identify the things no one expects to happen and provide support as a catalyst to learn.

In the past, you've discussed that the way the workplace is constructed has a critical influence on employee productivity, passion, and innovation. How can businesses appropriately construct a workplace that fosters these behaviors?

There are many components that make up a high-impact workplace. One is helping workers understand what the company's highest performance challenges are. Most workers are siloed and see only their piece of the puzzle. Too often, they have no sense of how they make a difference. We should help them focus on the challenges they can address, solicit their ideas and suggestions, and encourage them to bring forward creative approaches to solving those challenges. In many environments, work is so highly structured that experimentation may be viewed as high risk, and if a worker does something different it may cause a ripple effect. This can lead workers to mitigate experimenting with new ideas because they fear the fallout. I'm intrigued with the idea of creating experimentation platforms—systems and tools that create environments where risk is minimized. You will fail more than you succeed, but risk can be contained so it won't have profound impact. We need to encourage workers to experiment, which increases learning. Another high-impact component is technology itself. Increasingly, as workers interact with one another to solve problems, the solutions can be captured as the interaction occurs to create interesting knowledge repositories, helping subsequent workers quickly find solutions to problems.

Do you think the Internet has changed the way we think, or does it have the potential to?

Google has affected us in many ways. But I think this question depends on the person and his or her context. I agree that the technology has significantly diminished our need to know a specific fact because the answer is at our fingertips. On one level, we are now free to focus on creativity and imagination rather than memorizing facts. I think the more negative effect of the technology, however, is that it has inadvertently created an attention deficit effect in terms of multitasking. We get distracted by one link, which leads to another, and before you know it we've consumed an hour and haven't accomplished anything. It's important for all of us to have a sense of what our priorities are and how we use the technology to support those priorities rather than undermine us and distract us.

Artificial Intelligence

Labor-intensive jobs are not the only ones affected by automation. With the advent of artificial intelligence, jobs that we consider to be highly skilled knowledge work will also encounter massive dislocation and transformation.

AI is the science and engineering of making machines intelligent. It's easy to think of AI as a recent technology, but in fact, it has been with us since the 1950s, when Alan Turing wrote about a test he devised to determine if a machine could be considered intelligent. Turing's 1950 paper, "Computing Machinery and Intelligence," included a test in which a human and computer were questioned by an interrogator that didn't know which was which (Hodges n.d.). Turing's test argued that if the interrogator couldn't distinguish who was responding—either the computer or the human—then the computer could be called intelligent.

Since Turing's time, people have diligently worked to make machines intelligent. The general approach to AI has been predicated on how well a machine can solve a particular problem, not on its ability to look at a problem from multiple angles, apply reasoning, and determine the best outcome on its own. To be truly intelligent, the machine needs to receive information, detect patterns, adapt to changing conditions, predict outcomes, and then synthesize that information to determine how to act. This series of events—finding, learning, and behaving—is what human brains do by default. Some researchers argue that until a machine can accomplish that, it is not truly intelligent.

Since the advent of the personal computer, we have been programming machines to solve specific problems or do specific things. With AI, we can enable machines to go beyond being "dumb terminals" and operate with the same principles used by our brains. Jeff Hawkins, co-founder of Numenta, a company working on AI and machine learning, says that "the hallmark of intelligence is extreme flexibility, not the ability to solve a particular problem. The number of things humans can learn to do and the problems we can solve are vast. To build intelligent machines with equal flexibility we need to replicate the brain's principles in software." In his opinion:

> The goal of AI is not to create a human clone, or for us to become slaves of machines—that's the dark side of AI we often see in science fiction. Instead the promise of AI revolves around speed, efficiency, and the ability for an intelligent machine to crunch large sets of data to quickly uncover patterns. Think about how fast a machine can scan thousands of photos to find the three or four that have a correlation versus how long it would take a human to do that. Additionally, intelligent machines will not be burdened by human needs such as fatigue, hunger, and desire. (Hawkins and Dubinsky 2016).

Today, the main job of a human working with a machine is to establish the machine's requirements. However, this raises the question, will we always need to set the requirements for the machine? As computers get better at everything, they may eventually outpace human capability simply because humans are error prone and we've designed computers not to be. AI becomes most useful for us when it can set its own requirements. To achievethis, machines will need to learn on their own, without programming from us.

The biggest tech companies working in AI, such as Microsoft, Google, Uber, Amazon, and Facebook, see AI as a primary technology driving the next era of computing. According to the analysis firm Quid, AI has attracted more than $18 billion in investments since 2009 (Kelly 2016). In 2014 alone, more than $2 billion was invested in 322 companies with AI-like technology. Amazon has more than 1,000 employees dedicated to voice-recognition technology.

Google has emerged as a leading innovator in the AI technology space because it has pivoted to establish AI as a foundation for every service it offers, from search to driverless cars. Google's CEO, Sundar Pichai, sees the world moving from mobile first to AI first. "It makes sense that computing will be there in the context of what you are doing. You are trying to go about your day, and things are there to help you. You'll be able to do that because there are more intelligent devices" (Helft 2016).

A key element in Google's strategy is to enable machines to learn on their own without human intervention. In 2014, Google acquired DeepMind, an early-stage AI company that combines machine learning and neuroscience to build learning algorithms (Gayomali 2014). DeepMind's first area of focus was to build AI software that could teach itself to play video games. The important aspect of the technology was enabling the AI to learn how to play, as opposed to teaching it to play a specific game. The AI taught itself to play a 1980s game called Breakout, based on Pong. At first, its play was random, but it quickly learned how to increase itsscore, reaching the maximum score level after only an hour of play. After two hours of play, it started discovering loopholes in the game that millions of human players never uncovered. The AI did this on its own, with no human intervention.

We are decades away from machines mimicking the brain enough for them to be as creative, innovative, or even be able to apply reasoning and critical thinking on their own, but there is useful AI technology available to us now, including conversational assistants.

Conversational Assistants (Chatbots)

Both Google and Apple are beginning to infuse more AI-enabled technology into their smartphone operating systems with notifications and "head up" information. This first wave of AI is focused on developing useful conversational assistants, sometimes referred to as bots or, more recently, chatbots. By leveraging advances in search, image recognition, voice recognition, and translation technologies, the first wave of conversational assistants represents a narrow but useful application of AI by helping you complete mundane tasks.

At its most basic, you speak to a conversational assistant, which uses algorithms to process data and respond accordingly, and then applies natural language processing to speak back to you. At this point in the evolution of the technology, however, it's quite easy to ask a question in a way that breaks the chatbot's ability to answer correctly.

With most current conversational assistants, the dialogue begins with the human starting the interaction by asking a simple question, such as, "What's the commute to work like this morning?" It's a relatively easy query to parse. A more difficult, yet more useful task, is for the assistant to proactively provide information, such as alerting you about the weather three days ahead in San Francisco because you're scheduled to travel there on a business trip. Or, initiate a two-way conversation with you, such as, "It's your wife's birthday next week. Should you order her favorite flowers?" You then respond with, "Ah, thank you. Can you show me options?" Within seconds, the assistant displays different options for roses. You tap a choice, and then seconds later you're informed the flowers will be delivered on your wife's birthday.

Your conversational assistant can parse that data to ensure a successful transaction because it knows who your wife is, where she lives, her favorite flower, your favorite florist, and your preferred payment method. This is representative of a natural conversation that reduces the need for you to take the time to engage in those tasks. In under a minute you and your assistant have performed a task that would probably take significantly more time to complete alone.

For AI to provide helpful assistance such as this, complex software needs to determine myriad details about you and your context in real time. Technology companies are investing heavily in conversational assistance because they see interactions like this as an expectation, as well as a competitive advantage. The key is for the assistant to anticipate your intent. Soon, from the moment you're alert to the moment you close your eyes, your smartphone will serve as an active agent in your life, anticipating and answering questions, and completing tasks before you think about them. AI, in this form, is becoming one of the most consequential technological evolutions since the rise of the smartphone. Enabling a device to free us of the mundane unlocks the potential to boost our effectiveness to a whole new level.

The AI-Powered Knowledge Graph

A natural extension of a smart assistant is one that provides knowledge on demand. Researchers are working on specialized AI assistants that can build comprehensive graphs of knowledge throughout an entity, such as a corporation, and enable relationships to be built between the workforce and the information contained in the knowledge graph. Specialized AI will provide customized knowledge-based services to the individual worker by sifting through information to deliver only the information relevant to the worker's needs. The knowledge graph will drive quicker speed-to-readiness for new workers by providing onboarding assistance and connecting the new hire to resources applicable to his or her specific roles. Think of this specialized AI assistant as a virtual coach for employees.

Unified Interfaces

The most desirable application will be a single user interface that reduces the complexity in how you employ a conversational assistant to perform truly useful tasks. This means your assistant will be with you regardless of where you are or which device you're accessing. Currently, if you have your iPhone with you, Siri is your assistant. If you grab an Android device, you have Google's assistant; and if you walk into your living room, you may have Amazon's Alexa to help out. What you really want is one assistant that knows you, has learned your preferences, knows how to assist, and is transportable across devices. You don't want to have to start over with an assistant on a different service provider.

Business Applications

Advances in voice recognition and natural language processing are great for personal consumption, but business applications for AI are also accelerating across several enterprise industries:

- Automated legal assistants conduct case research, saving law firms thousands of hours of human paralegal time.
- Smart medical assistants help determine patient diagnoses, giving doctors and nurses more time to connect with their patients.
- Data assistants save time organizing and responding to unstructured data such as procurement requests, claims processing, contracts, and other types of business correspondence flowing in from customers, suppliers, and partners.

Tasks like these rely on automated assistants to learn patterns through repeated processing. If the assistant isn't confident about its decision, it can request assistance from a human. This requires highly complex AI engines so that the assistant can parse the right data to make the right decisions. The impact will transform business productivity by automating almost all clerical work and streamlining or eliminating almost all administrative tasks.

The next evolution of AI is already moving from automated agents assisting us with work tasks to replacing humans in more complex situations. Uber has automated every aspect of its business, including billing, navigation, and communication systems, except for one important component: the human driver. However, there are initiatives underway to develop fleets of driverless cars that will speed alongside human-driven vehicles on our roads and freeways.

Learning Applications

What does all this mean for your learning organization? AI is far from being a mainstream technology right now, but consider how you can begin thinking about how to leverage it in the learning experiences your team designs and delivers. For example:

- augmenting synchronous training with automated training agents that provide the latest thoughts on leadership principles to executives on a recurring cadence; the training agents can even provide reinforcement for material learned in past classes
- using learning bots to curate content for specific subject matter areas, ensuring the material is up-to-date, links are active, and searches are filtered based on guidance from instructional designers; the bots can also provide auto-generated activities based on learner input (for example, after three completed learning activities, the learning bot creates a new activity factoring in learner progress
- evaluating assessments and providing alternative or additional learning opportunities based on learner scores
- conversing with learners on foundational concepts and principles, such as defining basic terms, assisting with pronunciation, and providing guidance through complex processes or procedures.

AI can also help your staff become more productive by:

- reducing time spent on content development, with automated assistants mining content, conducting searches, checking grammar and syntax, and building outlines
- assisting with curriculum management and evaluation of curriculum effectiveness by uncovering patterns across large data sets in a much shorter time than humans can
- creating automated notifications and reminders for event-based learning initiatives.

AI will become more mainstream as the technology advances and, more important, as developers create more usable AI applications. When Apple introduced the first iPhone in 2007, there was no app store. In fact, Steve Jobs didn't initially plan to allow third-party native apps. When Apple did create the app store, it transformed the iPhone ecosystem. The same will happen with AI applications once there's a platform for creating revenue, such as an app store.

For developers, this new marketplace will grow like the app stores did. In the next several years, there will be thousands of applications to leverage across various domains, including behaviors, subject matter specificity (how-to areas), and commerce. Although AI applications are still nascent for the learning organization, start preparing now:

- Ensure someone on your team is thinking about how AI can be scaled to increase productivity and provide new learning deliverables for your audiences. Developers are already creating interfaces that will simplify AI across a host of devices.
- Look at collective intelligence platforms that are emerging and identify strategies for integrating them into your overall content strategy. AI's best application for learning will be in auto-curating from large data stores. You can leverage this for customized product knowledge training across large inventories and individualized knowledge bases (based on context and customer service), and deploy content across multiple channels (mobile, tablet, desktop) without significant redesign work.
- Work with businesses in your organization to identify processes and problems that can benefit from AI. You will need to have the skills to create automated agents to support those businesses. Working on small pilots now can inform your overall learning content strategy.

AI brings a universe of capabilities to the business. This technology will have a massive affect on the workplace. Many are realizing, however, that less work does not have to mean less productivity. Technology enables people to do more in less time, but with that will always come other things to accomplish.

Ask the Expert: Gregory Abowd on Learning Technology in the Workplace

Gregory Abowd is a computer scientist best known for his work in ubiquitous computing, software engineering, and technologies for autism. Abowd's work has focused on human-computer interaction and its affect on the usability of interactive systems. According to Abowd, we've gone through four generations of computing in just the last 60 years:

1. *Mainframe computing*—Mainframe computers are "scarce-resource" systems, usually administered by government and large companies, and designed to handle high-volume input and critical business applications. Mainframes were popularized by IBM in the early 1960s and are still in use today; in fact, IBM released a new mainframe system in 2015 (Wikipedia 2016).
2. *Personal computing*—The personal computing revolution began in the 1960s, when manufacturers began designing computers for individual use, and became mainstream in the 1980s, when the number of people using their own computers surpassed the number using shared computers. The personal computing generation was defined by the "one person, one computer" principle.
3. *Ubiquitous computing*—Coined by Mark Weiser in 1988 at Xerox PARC, ubiquitous computing is described as "invisible, everywhere computing that does not live on a personal device of any sort, but is in the woodwork everywhere" (Weiser 1996).
4. *Collective computing*—Coined by Abowd (2016), collective computing heralds a new era of cooperation between humans and digital technology, primarily influenced by the confluence of three technologies:
    - the cloud, which is the maturation of distributed computing that provides near-limitless resources through network-enabled services
    - the crowd, which is the rise of social computing platforms connecting humans and enabling communication and curation of knowledge
    - the shroud, which is the emergence of the Internet of Things (IoT), creating a layer between every physical object (including humans) and the digital world.

How do you think the prevalence of technology affects workers and the workplace?

A good example of how technology affects the workplace is how personal navigation technology has evolved. You can easily find out how to get somewhere and how long it will take. On-demand personal navigation systems have revolutionized mobility.

Additionally, technology has enabled us to harness on-demand expertise. There's good and bad in this. The good is we have tons of information at our disposal to help us with whatever we're doing. The bad is what were previously mostly manual activities are now enhanced by technology and require the worker to have some understanding of that technology, adding additional complexity to what was once a manual job.

For example, the new automated Coca-Cola machines offer the consumer a great experience because it allows them to mix a single drink from a wide selection of flavors. However, maintaining and troubleshooting the machines is complex, requiring the worker to know how to program it if something goes wrong. This era reminds me of a famous quote from Alan Perlis, "Prolonged contact with the computer turns mathematicians into clerks, and vice versa" (Perlis 1982). At times, it seems the scientist ends up doing things they would have delegated to their staff. On the flip side, the staff ends up learning a lot more about technology simply because the computer was introduced. This is not a new phenomenon. It's accelerated because of the ubiquity of technology.

When you look at artificial intelligence, what do you see as its biggest positive, and its biggest negative?

Automation will relieve humans of performing risky and dangerous work. Machines will be closer to the danger, and instead of taking on the risk, humans will be in supervisory roles, away from the danger. The idea of knowledge work will become less tangible. Companies will always seek to reduce the requirement for humans to perform expensive processes when there is economic incentive to apply automation.

We are now living in an age of networks. What do you think this means for worker skill sets? What's the most important skill set for the worker of the future?

It's thought provoking to consider what skills will be the most valuable. I do believe there are basic skills with fluency in information technology that people will need to have. Today we take for granted that people can make a phone call or operate kitchen appliances. We learn a certain amount of technical savvy every day, which will now always be required.

As computing evolves, what do you foresee as the single biggest effect on the work world of tomorrow?

Individuals now have the ability to harness on-demand expertise that they do not have the proper training for. I can become my own doctor quickly through online medical advice. Technology provides the ability to get any information you need at anytime, anywhere.

We view AI in the short term helping workers with administrative roles in the workplace. What's your thinking on how intelligent these components can actually become? What's AI like five years from now? In 20 years?

Speech recognition in the last few years has made strides and is having a huge impact. These technologies don't always come about fast—it can still take a long time for a new technology to become ubiquitous. We know that everyone will be connected, but what is not certain is the best way to facilitate that connection. In the emerging world, wireless technology has leapfrogged traditional wired connections. In time, a network connection will be viewed as a utility, like electricity. People will just expect the network to be available.

If you were in a corporation's shoes, what would be your focus on the future of training?

As an educator, it's important to educate people to learn on their own. I teach programming to people who have no programming background. I don't teach a language that may be obsolete. I teach them to not be afraid of technology, and the process of how to pick up a new programming environment. I would advise the corporation to focus on teaching people how to be lifelong learners.

Virtual and Augmented Reality

For decades, trainers have designed simulations that immerse learners in real-world situations to convey key concepts and principles. Today, learning designers methodically apply various combinations of media formats to replicate learner contexts in the two-dimensional realm of digital learning. However, with the advent of virtual reality (VR) and augmented reality (AR), learning designers now have the power to blend both digital and physical worlds and place learners directly in the middle of the action to create truly immersive simulations.

The underlying technologies that drive both VR and AR go back many years, but the recent confluence of hardware devices, affordable computing power, and near limitless storage, has accelerated adoption. The key to understanding how to leverage VR and AR technologies is to first understand what they are and what they're good at doing.

Virtual Reality

VR has become a common term for almost any type of immersive simulation. Using VR, you can create a realistic virtual world with which learners can interact. A successful VR experience is one in which learners cannot easily tell the difference between what is real and what is not. Additionally, VR usually requires wearing a device such as a helmet or goggles similar to Facebook's Oculus Rift, HTC's Vive, Samsung's Gear, or Google's Daydream View and Cardboard.

Augmented Reality

AR combines virtual reality and the real world. With AR, you can interact with virtual content in the real world and can distinguish between the two.

Differences and Similarities Between VR and AR

VR and AR have the primary goal to immerse the user in a realistic experience; however, each technology does this in different ways (McKalin 2014). VR places the user in a completely isolated, simulated world removed from real life; AR projects a digital "layer" of content over real life, providing more context to what is seen. The content overlay can display supporting information about what the user is viewing, either through a smartphone or smart glasses, thereby constructing an entirely new digital interface over almost any real-world environment or object. As both technologies have evolved, VR has been used more in video games and social networking, whereas AR is being used more in education and training.

Since the early 1990s, both government and business have been experimenting with how to use VR and AR to increase efficiency and productivity, as well as reduce human error. Both have unique applications in learning contexts to increase performance. Because VR is more immersive, it can be used to replicate a complete environment. A unique advantage of VR is its ability to change environmental conditions based on user input, and for the user to receive immediate feedback from their input. VR is commonly used for:

- Field training for soldiers or flight training for pilots. Situations can be adapted to focus on problematic scenarios, narrowing in on the specific context or condition that might be troubling.
- Skilled trades such as welding, construction, and design. VR is a more cost-efficient way to train because practice doesn't have to occur on actual materials, and learners can repeat the tasks without replacing materials.
- Training for tasks that involve accuracy and speed. For example, in a distribution center a task may require a worker to load items onto a truck, effectively maximizing the space to load as many items as quickly as possible. Practicing how to appropriately load various items of differing sizes and weights allows them to easily try different loading scenarios to optimize their loading technique.
- Traditional medical therapy, including treatment of post-traumatic stress disorder (PTSD), and physical therapy to alleviate pain. Life-saving methods are being taught through VR simulations, including emergency room procedures (such as intubation and catheter replacements), preparing surgeons ahead of time to perform difficult tasks under stressful conditions.

Initial uses for AR have centered on performance support—providing how-to or support information during a work task. In the mid-1990s, Boeing experimented with AR to see if airplane technicians could complete wiring tasks in less time and with a higher rate of accuracy using an AR overlay versus technicians who did not use an overlay (Nash 1997). The AR system consisted of a headset and a small computer attached to the technician's waist. Wiring an airplane is a laborious and complicated process, entailing thousands of wire bundles with varying lengths that must be attached to the correct connector. Mistakes can cause a cascade of problems resulting in delays and cost overruns. The AR system used cameras on the headset to identify the workspace and then guided the technician to successfully attach the wire bundle to the correct connectors. Results showed that technicians aided by AR were able to complete the wiring process 20-50 percent faster than technicians who were not accessing the AR system. Because the technicians were not constantly referring to a paper parts list, their wiring attachment rate was also more accurate.

Another use of AR includes surveillance. For example, border agents use AR for border security. When a truck approaches a border crossing, surveillance cameras capture and process images from all angles, similar to cameras at traffic lights that snap photos of license plates. Analysis of the surveillance video taken of the truck can be instantly relayed to border agents: "Vehicle chassis is two inches lower than standard make. Search all compartments." (Belissent 2016).

Workers can use wearables to receive data while performing the work to assist them. For example, a public works engineer can receive data through her AR glasses to identify faulty infrastructure, determine the cause of the problem, and identify a fix. Or, imagine a first responder in a helicopter after a hurricane. He can get information about damaged buildings and infrastructure as well as the progress of rescue efforts on the ground. Wearing AR glasses, he would be able to see details of the buildings, what types of buildings they are, possible occupancy levels, and labels that indicate which buildings have already been searched and cleared.

Applications for Learning

Although it seems natural that learning would be the first market for VR and AR to disrupt, the formative years of the technology have seen it being used more in entertainment, marketing, and live events. The largest hurdles to adoption for learning organizations are the same as with almost any new technology.

Designing the Most Efficient User Experience

The user experience is key, especially with VR. Many users have complained about "cyber sickness" when engaging in VR experiences (Murphy 2015). This condition, similar to the woozy feeling some people get when on a ship in stormy waters, means that it's important to consider the ramifications of the experience on the user's physical state when designing an immersive learning experience—not a typical requirement in learning development. AR seems to have lower barriers to adoption from the user experience perspective, now that portable devices are ubiquitous. However, hands-free devices such as smart glasses or body-mounted displays, are still nascent technology and are not widely used.

Technology Constraints

Currently, VR technology is more mature than AR and is seemingly on a faster track to becoming more widely adopted. AR has significant challenges to overcome, including display technology, battery life, limited real-time processing capabilities, and calibration of the real-world physical environment with the digital overlay of information.

Content Development

VR and AR create more intuitive ways to interact with a device. This requires a different type of content development than that used by typical learning organizations. Similar to video games, immersive simulations provide wider fields of view, gestural interactions, and rich graphics. The design canvas is quite different from most e-learning programs. VR and AR aren't confined to the two-dimensional world, so experience designing for three dimensions is critical. Designing content as an overlay to the real world is also unique, and instructional designers will need to immerse themselves in the context of the environment for which they're designing and make affordances for changes in that environment.

Cost

As with any emerging technology, cost is a concern. Initially, VR headsets were expensive, ranging anywhere from $600 to $1,500 per unit. AR smart glasses such as Google Glass initially hit the market at about $1,500 each. As with smartphones, once the technology matures, price points will decline, thereby increasing adoption.

As the technology matures, however, more use cases for learning will become apparent. In fact, the use of VR and AR in learning will eventually become the standard tool for training because of authentic use cases and real-world applications. For example, guided AR overlays can equip less-skilled workers with the ability to perform complex repairs with limited training, reducing the time needed for out-of-context training. Think of the AR overlay as the virtual expert guiding the performer. VR's applications in medical, military, and engineering training are numerous, and it's estimated that the global training market for VR technology will be more than $9 billion by 2025 for military training alone (Bellini 2016). VR is also useful for the work teams creating learning experiences. In the VR experience, being present means remote workers can more easily connect to one another as if they were in the same room.

Ask the Expert: Enzo Silva on AR in Learning

Enzo Silva, a learning strategist for SAP, has been on the forefront of immersive learning technology for several years. Recently, his work has been focused at the intersection of learning and innovation. Silva provides a glimpse of practical applications for AR in learning.

Could you provide a good definition of AR in the context of learning?

Augmented reality is an enhanced perception of the world through multimedia digital content layered on top of reality. The most common layers of digital information available in AR technologies are video, graphics, audio, geolocation, and simple interactions, such as visiting URLs. By overlaying digital content on top of their current view of the world, learners can interact with information in rich, immersive contexts in location, on the job, and on demand. AR provides learning in a contextual environment that offers more authentic situations than what's available in physical reality.

How do you see AR affecting business?

AR has gained significant traction primarily due to the ubiquity of advanced mobile technology and the desire for more personalized experiences. Industries such as architecture, construction, manufacturing, education, and entertainment leverage AR to provide interactive content to help their users make more informed decisions. Large retailers, for instance, have been experimenting with AR to enable their customers to visualize what the rooms in their home might look like remodeled and refurnished using digital content layers for a personalized shopping experience.

What are some examples of AR in learning?

AR can provide learners with context through natural, real-world interactions. Imagine onboarding experiences that help new hires get acclimated to their surroundings through information overlays providing navigation cues, company facts, and department information. Other learning applications include those that enable safe practice for tasks where real-world failure has significant implications; surgical procedures, military training, and airplane piloting are all good examples where AR makes learning experiences more tangible, immediate, and immersive in real time. Other examples include simple information displays to provide richer meaning—a learner can simply point a smartphone at an object, such as a hardware device, an automobile, or art on a wall and display pertinent information over or next to the object.

Where do you see this technology best applied in learning?

Information is everywhere, and the amount of information and sources of information is exponentially growing. One of our biggest challenges as learning professionals is to not create more content, but instead create more meaningful contexts to help learners gain the skills they need. A key priority is to help learners make sense of the world around them, their careers, and their aspirations. AR, if used properly, can be instrumental in the learning process. You can provide learners with a layer of information that offers the content they need to perform a specific task, in the appropriate context without disruption, without the need to stop what they're doing or go elsewhere to find the information they need.

Context should become the primary work of the instructional designer in this space, not creating more content.

What are some of the tools learning professionals can use to get started?

There are several common tools for AR development, for example:

- software development kits such as Vuforia, AR Toolkit, and Wikitude
- game development toolkits such as Unity and Unreal Engine
- 3-D modeling tools such as Blender
- Google Tango
- AR editors such as Aurasma, Gamar, Tamar, Zappar, and buildAR.

What are the downsides to AR?

Caution in proper design and implementation is necessary with AR, just as with any other technology applied to learning. You don't want to use a technology like AR just for the sake of using it—you want to apply it in the appropriate learning situations. Currently, AR requires significant proprietary technology (such as software development kits, end-user applications, and hardware), and disseminating this type of technology in a learning organization can be a challenge. A lot of the development done in the AR field relies on closed systems and a very limited pool of technical expertise. Implementing AR in a learning context, in some ways, is not dissimilar to attempting to make instructor-led materials work in an e-learning context. It's important to use AR for the affordances it provides (immersion, interaction, and intuitiveness), not simply because you can transform a flat, 2-D interface into a more immersive 3-D environment. Let the learning needs drive the experience.

What skill sets are required to develop AR experiences?

Skills needed to design and develop AR experiences vary immensely, potentially converting your organization into a cross between a learning design team and a software company. You need meaningful interaction design and thoughtful performance optimization to create successful AR. You can delve into simple AR experiences by having an instructional designer use a visual web-based AR editor that augments a physical object, such as a company logo or a graphic with videos, links, and other simple forms of interaction. However, to create AR experiences that go beyond a basic set of interactions you will need a team of media designers, 3-D animators, and storytellers to create rich, visual, immersive experiences. Technical acumen is necessary as well, including applications engineering, HTML5 programming knowledge, user interface design, systems integration experience, and potentially, data scientists capable of extracting analytics that help determine the efficacy of the learning experience.

What does it cost to develop AR experiences?

Organizations can start with readily available, free, or "freemium" applications to develop basic AR experiences. It can be difficult to estimate cost simply because of the sheer number of technologies involved, the software licensing, and the amount and type of content, which can range from custom video to graphics and more. It may be more valuable to partner with a third-party that has expertise in AR as you get started.

Key Takeaways

Historically, from the agricultural revolution that brought us the farming economy, to the industrial revolution that moved workers to factories, to today's era of globalization, the number of available jobs has always increased. The pace at which workplace automation is occurring is redefining work and reshaping the skills we need to learn in almost real time. This raises two questions: What skills are necessary, and which skills should you focus on for humans and machines? From now on, we will be bombarded with data and information–granular and hyper-individualized data. The ability to extract meaning from that data is a key skill, one in which humans and machines will need to work together.

Companies will move to strike a balance between finding the jobs at which machines excel, identifying the jobs more suited to humans, and then reinventing processes to leverage the maximum output of both machines and humans working together at what they each do best. Humans, for example will always be better at product development and strategy, whereas robots will do the heavy lifting.

We're at a critical juncture in determining how to train workers to prepare them for today's jobs, as well as future-proof for the needs of tomorrow. Our present-day learning method emanated from the British Empire more than 300 years ago—teach the masses to perform the same jobs in the same way over and over, creating a system of worker bees. Just like today, people needed to share reliable, credible information with one another across the globe. The difference is they didn't have computers or the Internet. Even though we have computers capable of delivering unheard of amounts of information in real time, training then as now is still too programmatic and generalized.

We're getting to the point where knowing will not be as important as it once was because software, robots, and other intelligent machinery will learn and perform critical tasks. Finding information and making decisions will be the primary skills humans need. The emerging network of robots, intelligent machines, and connected devices working alone and together in both structured and nonstructured environments poses a significant challenge for learning teams.

Technology is far from new for most workers. It has always shaped how we work and what jobs we do, but there is a palpable fear in some circles that automation will bring about mass unemployment and change the very structure and idea of work itself. The advances such as the ones discussed in this chapter show us that there is almost no limit to what humans are capable of when it comes to technology and science. Now the question before us is how will we balance our notion of work and the satisfaction it brings with our unceasing drive to invent new methods to make our lives better? Furthermore, how will we adapt to a world where the tasks we're used to doing will often be performed by machines? What is left for humans in a post-work world? One thing is certain: For the rest of time, we will balance our careers with machines. Consider these key points:

- We are entering an era of intelligent, connected devices that are transforming and optimizing the workplace.
- Product development, IT, manufacturing, sales, marketing, supply chain, and service are all functions affected by workplace automation.

- Immersive experiences will become the standard for delivering the most authentic learning.

This evolution calls for a strategic shift in learning strategies in two key areas:

1. What do learners (machines and humans) need to know now to perform?
2. What do they need to access to know more at any given moment?

In the end, workplace automation should be more about humans than machines. That's a choice we can still make.

2

The Cloud

"If someone asks me what cloud computing is, I try not to get bogged down with definitions. I tell them that, simply put, cloud computing is a better way to run your business."
—Mark Benioff, CEO, Salesforce.com

TONY GARRETT MANAGES A LEARNING DESIGN TEAM for a Fortune 100 company. His 13-person team is completely remote, located across the world, and consists of both full-time employees and contractors. This distributed employee model is becoming more common with the advent of cloud technologies: a combination of information technology services in which resources are accessed from the Internet rather than a direct connection to a server (Investopedia 2011a). Tony's team collaborates, conducts meetings, and manages their work as if they were all in the same office. Virtual work is not a new phenomenon, and most large enterprises support some form of it. But for entire teams to work all virtual, all the time, the cloud offers multiple services that enable productivity, collaboration, sharing, and the feeling of being connected to co-workers.

The cloud offers both risk and opportunity for Tony and his team as they navigate this new work model. The all-remote team must create and maintain a collaborative work model that enables brainstorming, information sharing, and a sense of shared purpose. Another risk is the ability for the team to proactively identify and then overcome the feeling of isolation that may occur in people who work alone. The risks for Tony as a manager include effectively leading a remote team, setting clear expectations, and monitoring the work product. However, opportunities for both the team and the business include the power and capability to leverage a broad array of cloud-based technologies to deliver learning experiences to audiences at a quick pace, while also ensuring that the quality of the deliverables meets the team's requirements. Another opportunity is to establish a work environment that implements technology and human capability to focus on what matters most: creating a fulfilling work environment that facilitates productivity, purpose, and creativity, regardless of location or physical togetherness.

Tony's team is truly a human cloud: remote workers, spread all over the world, who come together through the use of cloud-based services to accomplish their work. The cloud is forcing changes in business, while at the same time providing solutions and improving cost and time efficiency over older work models of designing and delivering learning experiences.

***

The Shift to the Cloud

The cloud itself is far from new; we use cloud-based services every day both at work and home. If you've used Sharepoint in your office or streamed a song on your smartphone, you've accessed the cloud. Technologist Kevin Kelly (2016) describes the cloud as "simply a continuous stream of data. Everything flows in and out of it as you work: your text messages, emails, documents, and media." Now, instead of the computer as the container for data, the cloud is now the container for our digital life, without the storage restrictions of the computer.

The Genesis of the Cloud

What we see as the cloud today didn't happen overnight. The cloud evolved from three key switching phases:

- **Phase 1:** switching content from paper to digital to the Internet
- **Phase 2:** switching applications from desktop-centric to Internet-centric

- **Phase 3:** switching infrastructure from traditional data centers and individual enterprises to Internet delivery.

Initially, security was the largest barrier to adoption within the enterprise. However, cost benefits, economies of scale, reduced management overhead, and the ability to temporarily acquire IT resources accelerated adoption. As cloud providers have produced better pricing models, process automation, and stability and security, adoption has accelerated even quicker in recent years. Companies that decided to build their own clouds started with the physical infrastructure layer, then added storage and computing utilities and platforms, and then application services. Cloud adoption accelerated because of these key value drivers:

- **Simplicity**—Easy-to-use interfaces enabled a strong focus on prototyping business ideas instead of worrying about complex technical operations and system administration.
- **Customization**—Programmatic and portable access to IT infrastructure and application programming interfaces (APIs) drove product and service development.
- **Affordability**—Pay-as-you-go cloud services reduced barriers to entry with little to no capital expenses for IT infrastructure required.

The Cloud Evolution

Cloud computing will continue this path to more automation and ease of use, allowing anyone to create and consume apps and services for the cloud. Cloud computing differs from other historical IT models; it focuses primarily on services, rather than technology. Within the next few years, traditional IT work silos—such as storage experts being organized on one team and network experts on another team—will erode as more virtual and cloud-based IT services become the standard, requiring cross-team expertise. Removing these silos enables IT professionals to have a more business-centric view of the overall technology stack. Increasingly, chief information officers (CIOs) are being looked at to provide IT strategies and budgets that produce a deeper alignment to the overall business imperatives. This includes a move away from managing infrastructure as just capital expenditure and toward increased integration of cloud-based subscription models. A key component in the decision to move away from on-site infrastructure is scalability: The expense of additional infrastructure will no longer justify the benefits, especially if services ever need to be scaled down. As IT continues outsourcing more services to the cloud, the idea of company-owned and operated data centers becomes less common, and as cloud technology evolves, the advantages to the enterprise become more apparent.

Flexibility and Massive Scalability

Based on business demands, it's relatively easy to scale cloud-based services, because you're accessing remote server farms and do not have to integrate expensive hardware. This type of operational agility—the organization's ability to quickly provide services and support aligned to business needs—enables IT to more proactively partner with the business to support its needs (Sull 2009).

Disaster Recovery

Too many organizations suffer productivity and efficiency deficits due to loss or mishandling of data and wrong or out-of-date backups. Cloud services can enable data protection policies that don't strain internal resources or require additional internal overhead.

Security

Employees inevitably lose or damage computing devices that have persistent connections to the network, exposing the company to risk. Because data are stored in the cloud instead of on specific devices, data loss is mitigated and devices can be remotely managed.

Subscription Model

Data centers require significant investments in hardware, software, administration, and energy. Cloud service subscription models reduce the need to acquire and manage hardware. However, there are significant disadvantages to consider.

Downtime

Regardless of service provider promises, outages happen. A connection to the Internet is required, which means that connection must be live for cloud services to be available. In addition, cloud services are still driven by hardware (CPUs, memory, storage, and software), which inevitably will fail. You must fully understand which services can be delayed or stopped completely if your cloud provider has a service outage.

Privacy and Security

All corporate data should be viewed as sensitive. With cloud computing, any Internet-enabled device could potentially access your data. By outsourcing cloud services, you're entrusting your data to the service provider. No system is perfectly secure, so your IT organization should conduct a complete risk analysis of the service provider to minimize privacy and security-related issues:

- Identify any procedures the service provider has in place for network protection.
- Determine how to extend these security measures to the supported devices that connect to the service.
- Limit data access based on user roles.
- Determine how data assets are secured, and if and how the provider monitors the system to mitigate risk.
- Investigate if the provider can conform to your internal IT compliance controls.
- Ask if the provider has a notification policy for when unexpected situations occur.

Less Control Over Infrastructure

For some, this is a disadvantage in that you may not have local access to the server to administer or modify its functionality. This includes application management, data management to some degree, and the server's services.

Support Versus On-Site IT

You may find yourself in situations in which you're reliant on the service provider's support over your internal IT organization. During the initial service provider analysis, it's important to probe service level agreements and seek other customer feedback as well to determine your comfort level with their support capabilities.

Cost

The inherent promise behind cloud services is the ability to scale up based on your needs. However, you also have to consider policies surrounding your ability to scale down when services are no longer needed or instances need to be stopped. The other key element of cost is the number of users accessing the services. If your enterprise is large, you may not want a pricing model that requires per-user access fees; instead, you would negotiate an enterprise license.

Learning and the Cloud

Before the cloud, learning organizations were required to partner with IT to purchase and maintain the costly infrastructure supporting learning delivery and management. In some situations, learning teams even managed their own technology. Cloud services have reduced, and in some cases completely eliminated, the need to maintain in-house hardware and software for the learning organization, reducing the complexity and resources required to manage the infrastructure. Because cloud technology can be implemented and scaled quickly and affordably, adoption has been swift, bringing wholesale restructuring to IT organizations. As corporations begin adopting cloud-based services, learning teams will be able to expand their current capabilities to create, deliver, and manage learning experiences with more productivity and efficiency than ever before.

Historically, learning experiences have been created based on facilitator-led learning, including printed and electronic materials for classroom training. Computer-based training, e-learning, and distance learning emerged to enable asynchronous delivery models. The creation and distribution of learning content, however, centered squarely on the learning organization, which determined when and where learning occurred. This delivery system is costly and time-consuming, preventing the learning organization from delivering learning to their audiences in an affordable, relevant, and timely manner. As cloud technologies emerged and distribution channels broadened, the very idea of how a worker accesses learning was transformed, putting more control in the learners' hands and giving them more power over how and when they access and consume learning content. The cloud also allows learners to make decisions on what type of content offers the most value. If your content fails to meet their needs, they can easily find other resources. With this in mind, you may need to focus on more relevant and meaningful content, reconsidering how you go to market with your training. By leveraging cloud technology, you can accelerate your training deliverables by:

- enabling cross-platform learning development without requiring internal resources to manage the technology infrastructure; this results in the learning team focusing on their core competency, which is creating effective learning
- decentralizing content, enabling curation, expanding collaboration opportunities, and providing the ability to add, modify, and delete content quickly; by moving content to the cloud and linking and connecting to other content repositories, you can offer complete search and discovery capabilities across all content, aiding learners in finding solutions quickly
- providing up-to-date content with broader distribution capabilities; content updates are instantaneous across all linked channels
- integrating social elements to establish dialogue between learners and the learning organization
- quickly scaling up or down to meet audience needs; it's much easier to bring virtual machines online than to buy new hardware

- optimizing internal processes; for example, moving content off employee computers to a centralized data store increases workflow efficiency and ensures content is always backed up.

As software and services move completely to the cloud, it's now relatively simple to show that every phase of the learning design and delivery process can occur in the cloud. This affects the learning business in three primary areas: content development, work effort, and business intelligence.

Content Development

Teams create, review, and share in one content store. This is the key to achieving more iterative design practices, enabling the team to release, iterate, revise, and release, which is similar to Agile software development. Learning content is stored in one place, increasing the team's productivity, and team work flows—including review, editing, and testing—occur in one place, reducing miscommunication and increasing efficiency.

Distributed Work Effort

Continuous work can occur on global teams without the challenge of version control. Anyone who is granted permission to work on a project can securely collaborate, review, and contribute. Real-time collaboration streamlines team communication, issue resolution, and project management.

Business Intelligence

Data collection is easier because all work flows are in a central repository. Learning content is less costly to maintain, is always backed up, and easier to search.

Key Challenges

Before moving to the cloud, it's important to examine the following key challenges for learning organizations as they integrate cloud services into their business. To be successful in transitioning, you need a disciplined, systematic cloud strategy—the cloud is less about a bundle of technologies and more about an integration of technology, process, and team capabilities.

Challenge: How do you create a pragmatic cloud strategy rooted in the definition and delivery of learning services linked to business outcomes? How do you envision cloud-based services supporting your organization in its overall mission? What are the differentiating factors based on how you get the work done now?

Recommendation: Identify the cloud services you will offer or procure, and ensure they're aligned to business needs. For example, if you need to support a mobile learning strategy, make sure the technology you're considering accommodates that delivery channel and that the complexity of using technology to design and deliver training is realistic based on your team's capabilities.

Challenge: How do you gain perspective on initiating and integrating cloud services in the types of learning deliverables you create? What about the cloud changes how you design and deliver learning? Do cloud-based services alter the learning services you can offer?

Recommendation: Identify the internal work flows and processes that will be affected by the cloud services you're considering. Be sure to map all your applications and workloads to those corresponding services. Gain a deeper understanding of what effect the new cloud services can have on how your team designs learning.

Challenge: How do you strike the right balance between cloud technology and its influence on your team?

Recommendation: Too many leaders put a primary focus on technology when transitioning to the cloud. For your team to extract the maximum value from cloud services, address these three steps early in your decision-making process:

1. Identify how to achieve business outcomes by moving to the cloud. Be realistic and pragmatic, mapping a multiyear strategy. You won't be able to realize all outcomes quickly.
2. Document the cost and operational efficiencies of making the transition to the cloud. Compare this to how you do the job now.

3. Identify where you gain more agility across the design and development life cycle, showing where you foresee vulnerabilities.

Challenge: Once you've moved to the cloud, how will you analyze the impact cloud services have on your process? Has the move to the cloud made your team more efficient, more reliable, and capable of delivering the quality your business and audiences expect?

Recommendation: To derive a true sense of the value cloud services bring your learning business, track these three areas closely:

1. **Automation**—Define the processes, if any, that have been refined or automated. Eliminating manual effort and repetitive tasks when possible is key to minimize redundancy. To get this information, analyze all processes affected by cloud services; you may find that some reach beyond your learning organization. During this analysis, consider how you can refine or eliminate inefficient processes and target manual repetitive processes to be automated.
2. **Governance**—Identify governance processes for the cloud services that support the learning services you offer; this is necessary to show evidence of cost-effectiveness and efficiencies. Start by defining the infrastructure and application components that make up the core services you need to drive the business from a learning perspective. Your goal is to require only the minimal elements you need to effectively deliver services, such as computing power, storage needs, operating system support, and network reach.
3. **Operations**—Think ahead to how you will monitor and maintain the environment. Identify the existing processes, such as those you have for your existing LMS support, and determine which ones may not work in the new environment. Design an architecture and document the new processes needed to address reactive and proactive operational challenges.

Plan for cloud services that consist of technology and application elements, not just the individual technology components. The cloud services you end up with will be sourced from internal IT teams, your own team, and external vendors providing private or public services. As the cloud evolves, most organizations will use hybrid cloud deployments based on business complexity. Be sure to identify how the learning services you offer add value. Ask yourself:

- Which of my audiences will consume the cloud services?
- What services do these audiences want or need? Which should I offer first?
- When will audiences normally use the service? For example, are there peak periods?

It's critical to take a systematic approach as you craft your cloud strategy. It's also important to work with internal and external stakeholders to achieve a broad perspective on the impact the cloud has on your learning services. As you transition to more interoperable technology and processes to drive the business imperatives, building relationships will become a core component in successful outcomes. Learning organizations are no longer on their own to determine their overall operational strategy from a cloud perspective. You will increase the probability of success if you apply rigor in examining your transition to the cloud from three perspectives—people, process, and technology—instead of jumping straight to technology.

Cloud Platforms and Services

A learning platform can be defined as the technology systems and processes you require to deliver learning experiences. The primary platform most enterprise learning organizations use is a learning management system. In some enterprises, the LMS is managed internally by the company's IT organization, and in others the LMS may be hosted off-site. In either model, the practical application of learning platforms is changing with the advent of cloud technology.

In some ways, we're creating new maps from old territory, and it's important to be acutely aware of the technological components and services that make up cloud-based platforms. In the current terrain, we look at the LMS as a pull system (a catalog of available courses, many of which are on demand). You may even have some push components that inform learners of what's available. As you evolve your strategy to support learning wherever and whenever it's needed, the idea of a centralized pull-mechanism seems quaint. The daily realities of the workforce dictate that learning has to be available across the spectrum of performance for it to be most meaningful and relevant. Your learning platform may also need to support multichannel publishing. In this model, you may no longer control the number and variety of networked devices that connect to your platform.

Furthermore, what is a platform in the context of what you do? Some consider mobile to be a platform, and even focus more specifically on Android or iOS operating systems. Or you may simply define the LMS as your platform. How you define your platform and enable the products and services needed for support is critical to your overall learning strategy. How do your learners access the learning content you create? How do you want them to access it? With a multichannel learning strategy, you may discover that one platform won't deliver everything you need.

The LMS brought a centralized place to access learning. It afforded learning teams a one-stop shop to deliver learning plans, courses, and guidance mechanisms so learners would know what they needed to complete. It helped track activity, attendance, and the completion of required courseware. Today's workforce is becoming more mobile and decentralized, and is interacting with more complex workplace situations. Treating learning as an event on an LMS platform is no longer sustainable to ensure relevance across learning needs.

Cloud technology has the potential to redefine the idea of a centralized learning platform. The idea of the LMS is shifting to serve as the data store for learner activity, not necessarily the central focus for the entire learning experience. The idea of content curation and deep linking between content sources means that it will be increasingly difficult for you to store all your learning content on one platform. Regardless of how the learner connects to the learning experience, you need to gather data that inform you of the learning effectiveness and completion requirements (if any).

To understand how best to leverage cloud technology in your learning organization, let's look at the classic definition of the cloud. The cloud consists of a series of components you can leverage across your organization to achieve various goals as they apply to content creation and distribution. The formal definition of cloud computing comes from the National Institute of Standards and Technology (2010) and states that "cloud computing is a model for enabling convenient, on-demand network access to a shared pool of configurable computing resources (networks, servers, storage, applications, and services) that can be rapidly provisioned and released with minimal management effort or service provider interaction."

As cloud technology has advanced, three primary types of cloud services have emerged: SaaS, PaaS, and IaaS.

Software as a Service (SaaS)

SaaS is a software licensing model in which access to the software is provided on a subscription basis, with the software being located on external servers rather than servers located in-house (Investopedia 2011b). SaaS prevents users from needing to install software on their local hard drives because it's usually accessed through a web browser. Common learning uses for SaaS include online content authoring tools, such as those provided by Adobe, Lectora, and Composica.

Platforms as a Service (PaaS)

PaaS are customizable platforms that develop, run, and manage applications without needing to build and maintain infrastructure. PaaS technologies are much more affordable than the cost of installing your own infrastructure. Vijay Sankaran, the CTO of Ford Motor Company, stated in a *Wall Street Journal* interview, that PaaS technology can replace "things that were previously done manually [and create] repeatable and automated [processes]" (Rosenbush and Boulton 2013). Cloud-based LMSs are an example of PaaS and provide what is often referred to as a platform-driven scale effect. With this type of technology, business can experiment with new services much faster than ever before.

A classic example of the platform-driven scale effect is Amazon Web Services (AWS). By creating an AWS account, a person with little more than an idea can have a service or product up and running in a matter of days with minimal financial outlay. Instead of buying or leasing expensive hardware and data services, products can be tested in shorter periods of time and modified on the fly based on user feedback. The effect scaffolds across all business functions globally.

Infrastructure as a Service (IaaS)

IaaS delivers the cloud computing hardware and software that powers the entire stack: servers, storage, operating systems, and networks as a service. Generally, IaaS is thought of as distributed resources that enable dynamic scaling based on demand. For learning organizations, IaaS makes sense as your distribution models grow and provide new services. The main advantage is in situations in which scalability and provisioning requirements are unknown: You can determine the viability of services before making a significant investment in hardware, quickly scale based on unforeseen rapid growth, and provide immediate support to lines of business that may have urgent but temporary needs.

Understanding Your Platform Needs

The cloud doesn't describe a single thing; it's simply a general term for a variety of services. Although you may rely on your IT department to make technology decisions, it's important to understand the different aspects of cloud services so you can assess your situation and discover the components that will drive your strategy.

Learning organizations have a huge stake in the emergence of new technology across the enterprise. It's time to assert yourself when technology decisions are being made. To enable more timely and relevant learning, the platforms necessary to deliver the experiences should take center stage in your operational strategy. By making affordances for both the content strategy and the technology strategy, you'll go farther faster in realizing the learning business goals you have established. The cloud helps you transition from providing event-based learning to broadening your capability to deliver multiple, distributed experiences designed to reinforce performance at the right moment or provide the content on specific topics that learners need. As we know, not all learning in context is related to performance support. Providing support to assist learners in completing a task at the time they're actually doing the task is true performance support. This is an advantage in a truly adaptive learning model that leverages the platform across channels.

However, not all learning is related to tasks at hand. Your platform also needs to deliver learning experiences that help learners construct knowledge they can apply either outside their immediate context or in addition to prior knowledge they have. The cloud helps you implement the right technology in the right place at the right time to extend access to learning content anywhere your audience is or however they want to consume it.

Cloud-Based Learning Management

A learning management system will more than likely serve as a core component in your learning services cloud stack. Brent Schlenker, a learning technologist and cloud expert, identified three areas for learning teams to focus on when considering cloud-based platforms as a core component of your learning strategy: cost effectiveness, ease of use, and adoption considerations.

Cost Effectiveness

Besides new options in learning solutions, the most significant impact for learning is the cost savings and ease of use in the management of learning curriculum. Cloud-based LMSs provide the ability to control technology without deep involvement from IT. If you're familiar with older LMSs that required server installations and management by IT resources, you know how complex and frustrating that can be. The cloud takes that responsibility away from your internal IT department so they can focus on other business-critical needs. With the infrastructure centralized in one location, vendors can maximize their resources as well. This allows them to offer you an LMS solution at a significantly lower cost.

Ease of Use

Many who use a cloud-based LMS for the first time are shocked at the easy administration and course creation. This ease of use makes initial implementation faster, easier, and significantly lower in cost. Course creation is fast and easy as well, offering your less tech-savvy instructional designers the ability to create courses and blended learning solutions without assistance. This also introduces the opportunity to use nontraining resources for course creation and maintenance. The subject matter experts you need to create content can develop and maintain their own courses. While this may seem unwise to many traditional training professionals, it is a beneficial way to scale training offerings and broaden content, even introducing user-generated content and content curation.

Adoption Considerations

If you have existing content that was created with legacy technologies such as Flash or old versions of SCORM, you will need to factor the conversion of those courses into your transition plan. Even if the transition is a simple upload, you still need to consider delivery across multiple device types. Training developers didn't have to contend with this even 10 years ago. But today's device ecosystem introduces smaller screen sizes and new interface options such as touch screens. This changes how designers approach learning content design and larger blended learning solutions.

Cloud systems also open up a new world of options for instructional designers. The legacy of creating one-off training course solutions is now only a small part of what you can offer learners. In addition to the option of consuming content across multiple devices, the cloud also offers the ability to create more complex blended learning experiences. These include everything from self-paced online content as prework, to scheduled live online class segments, to microlearning content delivered as notifications to support spaced repetition.

Content Strategy

The cloud is both a blessing and a curse when it comes to content. In our daily lives, we encounter a pervasive flow of content, and we filter, sift, research, forage, correct, create, curate, save, archive, bookmark, schedule, publish, and so on throughout the day. If you or your team systematically engages in these tasks, then you are actively developing a content strategy. Learning professionals strive to create meaningful learning content in the most efficient manner possible, whether or not there is a content strategy in place. When it comes to formally creating a content strategy, ask yourself, "What is a content strategy? Do I already have one? If not, do I need one?" Rachel Lovinger (2007), a content strategy director at Publicis Sapient, a digital transformation consulting company, defines content strategy as "using words and data to create unambiguous content that supports meaningful, interactive experiences." Kristina Halvorson, founder and CEO of Brain Traffic, a content strategy consulting firm, further defines content strategy as "planning for the creation, publication, and governance of useful, usable content" (Kissane 2011). Although there's no prescribed content strategy that fits all learning teams, a good learning content strategy considers two key components: content organization and structure, and content development.

Content Organization and Structure

What structure, if any, does the content need? Will you need to create content taxonomies? Applying a taxonomy is the practice of adding metadata to organize content. For example, based on a line of business or subject matter. If you envision using content across channels, you may want to consider applying a taxonomy.

How do you organize and structure content to support your audiences? You want sustainable, flexible content that provides a meaningful learning experience. If your team works locally on their own hard drives creating content, ask why. The cloud is too convenient and easy to not use fully. At the very least, shared content repositories are critical.

Content Development

To create meaningful content, it's important to understand the learner's context. This may not be easy. Some audiences are large and geographically diverse, and have a broad array of performance capabilities. Providing what they need, when they need it, optimized for the device and situation at the time they access it is often referred to as human-centered design. Tim Brown, president and CEO of IDEO, an international design and consulting firm, and David Kelly, one of IDEO's founders, call human-centered design "design thinking," and define it as "a discipline that uses the designer's sensibility and methods to match people's needs with what is technologically feasible and what a viable business strategy can convert into customer value and market opportunity" (Brown 2008). Initially, design thinking was primarily used in product design, but as technology has evolved it's been applied to business problems as an approach to reduce the complexity inherent to software and hardware integration. Design thinking can be applied to many challenges across a spectrum of daily tasks, helping to expose multiple paths to a solution, and it contains elements useful for training design.

Empathy

Gaining empathy is the act of placing yourself in the workers' environment to discover their explicit and implicit needs, which will influence the learning solution. This involves three steps:

1. **Observing**—view learners' behavior in the context of their work.
2. **Engaging**—interact with and interview learners in scheduled and unscheduled encounters.
3. **Immersing**—experience what your learners experience.

Ideation

Also referred to as brainstorming, ideation brings creative and innovative ideas forward based on business and learning goals. Ideating is pivotal in the design process because generating multiple ideas helps shape the ultimate solution.

Prototyping

Building prototypes gets you closer to realizing what the solution will look like and how it will function. Prototypes for learning should be something a learner can interact with, even if it's not in the final format.

Understanding Learner Behavior

Adding these elements to your content strategy provides a richer understanding of the challenges your learners face. However, you cannot base your entire strategy on only the learners' context, which is the entire environment in which they interact, including its physical and digital systems. Their behavior shows how they engage with the environment and interact with its systems. User behavior with mobile devices, for example, differs from user behavior with a desktop computer. Even more granularly, a person will interact differently with a company-provided mobile device than their personal mobile device. It's impossible to predict behavior across various systems because people interact with them based on multiple variables (such as stress, convenience, location, and distraction), but you should make an effort to learn the following about your target audiences:

- How do they access learning content?
- Do they understand the technological devices used to access the learning content?
- What are their preferences with respect to the devices?
- Do they use more than one device?
- What do they believe they can accomplish after taking the training?
- What can they learn while on the job?

Try to understand your learners' actions, constraints, emotions, and conditions not only while working, but also while learning something new. Understanding their behavior is not simple, or an exact science, but it's critical in designing the appropriate learning experience and for informing your content strategy.

Many learning professionals tend to default to creating new content for each learning solution. In many situations, you may find that a significant amount of content is already available both internally and externally. This is where curation can play an important part in your content strategy. The challenge is in maintaining governance over content that you don't manage and providing the right amount of context for the learner. Simply listing links to other content resources without providing guidance on why the learner should access the content is probably not going to help them. Ultimately, you want to employ strategies that enable you to reduce your content management effort, while still providing access to credible, substantive content. Content curation is more like moderating a community than creating and delivering content that doesn't require consistent updates.

Your content organization should be grounded in determining a way to solve for these answers based on where you are now. An effective learning content strategy relies less on technology and more on the vocabulary of your learners. Speak their language first, and then let the other pieces fall into place.

Every solution you create should include the strategy behind why you're creating it, and the strategy behind making it accessible to those for whom you're creating. Use these two elements as the foundation of your content strategy, and then focus on how to leverage them in the most efficient way. The primary focus of this book is how technology transforms training, but content will always be king. Those who create content should be able to maximize the systems in which they work to create the best content in the most productive fashion.

Ask the Expert: Seth Malcolm on Cloud-Scaling L&D

Seth Malcolm is a senior operations program manager at Microsoft supporting IT, where he focuses on cloud technologies and enterprise networking services. An expert on IT infrastructure and its impact on the business, he discusses how cloud-based technologies affect modern learning organizations.

Everyone talks about the cloud and its impact on business, but can you share your perspective on how cloud technology affects modern learning organizations?

Think of the cloud as a delivery mechanism for services, content, and data. In its simplest definition, the cloud is a group of networked resources that are shared. Companies that harness the capabilities the cloud provides will differentiate themselves from their competition. In some industries, its effective use has already created leaders that are very difficult to catch.

What are the benefits of moving to the cloud for L&D?

The primary benefit of cloud computing for training (or any content delivery business) is the way the cloud scales, providing what we refer to as elastic computing capabilities. This allows global reach and the ability to serve millions of users without large capital investments and the risk that accompanies those investments. This scalability allows rapid development of new applications and portals that are critical to modern training programs. Additionally, the cloud abstracts the training provider from many of the IT-related tasks that used to be required for operating an online portal or system.

What are the disadvantages?

The only disadvantages are poor cloud adoption without the right architecture, and cloud-based applications can cost more than a legacy design over time.

When creating a learning strategy, what considerations should learning leaders give to cloud technology?

I think cloud computing should be a given, not a consideration, in any content-delivery business. Modern software design should be the first consideration; the speed of change in the cloud computing environment is swift. Having the right software architects who make good decisions now will profoundly impact the long-term success of an online portal or service.

What are the biggest components of cloud technologies that you see affecting the learning organization?

The integration of common identity/authentication systems improves the user experience significantly. Additionally, video delivery through cloud services is unmatched by legacy systems.

Case Study: Using the Cloud to Support Multichannel Learning

One of the largest public utility damage prevention companies is located in Indianapolis and serves all of North America. Its primary focus is to protect and maintain the infrastructure that provides public utility services. The majority of the infrastructure is underground and consists of the pipe, wire, and fiber optic cable used by the majority of public utilities and telecom companies in the United States. Its service requires a strong commitment to public safety, a driving factor behind its commitment to providing comprehensive safety compliance training programs for all employees. In fact, all employees undergo three weeks of classroom and field training when they first join the company.

A large percentage of employees are field technicians, often referred to as "locators" because they locate the underground utility infrastructure that needs repair. Locators leverage technology in their work flow, using hardware that detects the infrastructure with a high level of accuracy. They gather large amounts of data detecting and analyzing the infrastructure. In fact, they transmit more than 1.5 terabytes of data every day, including more than 7,000 photos. Employees can often be seen drilling roads, digging through rough terrain, scanning residential yards, or excavating underneath large public structures. They rarely work in an office, use technology to perform their work, and have many interactions with the general public. Gathering employees into a classroom for training not only presents a geographic challenge, but also costs the company money to pull workers off the job site. However, proper safety training is a key imperative for the company. All employees receive more than 40 hours of classroom training, leading to a field certification. The training program also includes extensive on-the-job training.

Historically, training was outsourced to a third-party vendor, but as the company grew to more than 7,500 employees, it needed to revise its reliance on classroom training. Because of the varying density of utility services in the United States (some locations are much busier than others), coordinating training efforts across locations became a logistical challenge. To be more responsive to its geographically dispersed employees, and to ensure content was relevant and up-to-date, the company decided to offer more on-demand training. E-learning would provide economy of scale, rapid content dissemination, and easier learning consumption for a mobile workforce. However, the company needed to leverage the cloud for this strategy to work. A cloud-based

LMS could track and report on training compliance and provide scalability and reliability.

Using a cloud-based LMS would allow the company to minimize use of its internal resources for delivering e-learning, because the LMS would be hosted through its provider's site. The training team would administer the system, including uploading course content, creating new courses, and storing trainee data, and communicate with the field securely through its web browser. The company realized these benefits:

- **Lower start-up costs and life-cycle cost predictability**—without physical software, the company was able to quickly launch a training site with little up-front cost. Administration was minimized because there were fewer requirements to integrate into the existing infrastructure. The training team was able to forecast its cost over time based on subscription fees, maintenance fees, and upgrade options. At all times, they knew their year-over-year commitment in cost and resources.
- **Broad accessibility and scalability**—the company gave its employees the ability to learn on-the-go, anywhere, and anytime, without needing to download and install anything. For course creation, the training team could use their tablets, mobile phones, and any other Internet-ready device to upload content.
- **Faster deployment**—when deploying an LMS, the biggest challenges are the initial setup, learning how to use the system, and preparing for rollout. After a comprehensive review of systems and services, the company established a relationship with its preferred provider in a manner of weeks and worked with the provider to strategize the integration, enabling the training team to focus more on content strategy and less on the technical aspects of deploying e-learning.
- **Security and reliability**—when using an external provider, it's critical to ensure data integrity and security. The company worked with its provider to ensure that its training team would not need to worry about data being lost or stolen. The company's IT organization reviewed the system based on its requirements for encryption and secure connections in and out of the system with full participation from the provider.

- **Unlimited storage**—although the cost of storage has decreased, it can still require a significant investment, especially if video-based training is being used. With the cloud, the company's data, including media, was uploaded directly to the LMS, requiring no server space allotment from IT. With backups, the fear of data loss due to corrupted hard drives or deleted files was also minimized.

Using a cloud-based LMS saved the company $1.5 million in training travel costs in the first three weeks of deployment, and continues to save significant amounts, while providing a multichannel learning platform. Their training now gets to the field quickly through mobile devices and adapts to meet their ever-changing needs. "Our employees like it and use it every day," says the manager of learning development. "It helps us make them safe and effective, and in the end, our customers are happy with what they are doing."

Key Takeaways

Some experts would say that the cloud has been the single largest disruptor in how content is distributed since the rise of the Internet. The cloud provides anytime, anywhere access to knowledge and understanding, enabling decentralized work teams to flourish. As long as the cost of computation continues to drop, the cloud will inevitable become your main learning platform. The cloud is the result of global expansion and ubiquitous computing. As our reliance on the network increases, the cloud will gradually displace other internal infrastructure, as well as the idea of owning hardware.

Mark Hurd, co-CEO of Oracle, has predicted that the cloud will "eat everything by 2025." In an interview with Julie Bort (2016) of *Business Insider*, he stated:

- Nearly all enterprise data will be stored in clouds.
- 80 percent of IT budgets will be spent on cloud services, as opposed to traditional systems.
- The number of corporate data centers will drop by 80 percent.
- CIOs will spend 80 percent of their budgets on innovation, not maintenance.

We're still in the early stages of our move to the cloud, but it's already obvious that it is a game changer for learning delivery. The cloud represents the largest migration in the history of the IT industry, and its influence has yet to be fully realized.

3

Mobile

"Mobile is eating the world."
—Benedict Evans, Andreessen Horowitz

SHARON GASKER'S ALARM STARTED PLAYING SOFT MUSIC at 4:30 a.m. She set it to wake her up extra early because this was day one at her new job. Sharon was now a marketing manager at the company of her dreams, a global provider of technology services. She wanted to get a head start on what would probably be a very busy day. She's eager, but not quite ready to jump out of bed yet, so she asked her alarm to give her just a few more minutes of shut eye.

After her quick snooze, she showered, grabbed her iPhone from her night-stand and headed out for breakfast. She saw a few notifications on her lock screen and swiped the first one. It was the HR department sending her links to an onboarding video and a quick, two-minute module welcoming her to the company. She could complete her initial onboarding while on the train to the office.

After she completed the two short training modules, she texted her new manager to let her know she already completed some of her required training before getting to work! In our new mobile-first world, Sharon is already a productive new hire before she is technically on the clock.

By providing orientation information to new hires before their first day, Sharon's company is not only helping her get a jump start on her new role, but also significantly reducing the time it takes new employees to become productive. Additionally, by creating and distributing a New Employee app, it's built a powerful recruiting experience it can use to target new talent.

***

On June 29, 2007, the iPhone was introduced and the world changed. The iPhone was the first mobile device not laden with bloatware and confusing hard keys and buttons. It was also seductive, begging to be touched. The user interface was intuitive. It responded to natural human gestures. The iPhone offered a great user experience, which previous mobile devices had failed to deliver. It opened a new realm of possibility for consumers and developers.

We have only begun to realize and understand the impact mobile devices and constant connectivity have on the workplace. For workers, the mobile device is insurance against not knowing what they need to know to do their job. From now on, we'll have access in our hands to information on a scale humanity has never known.

How has mobile technology affected the workplace? In just the few years since 2007, smartphones have become ubiquitous and touchscreens are now standard. Entirely new behavior patterns have emerged, as learners have become untethered. Learners now expect personalized learning experiences whenever they want and wherever they are.

Why Mobile?

Of the five factors, mobile is easily the one that has received the most attention over the last several years. It has rapidly become a central part of all our lives. Recent developments in mobile technology, such as easier app development, faster carrier connectivity, higher-resolution screens, and faster processors have quickly become the fuel driving the Age of Immediacy.

For years, most learning organizations have designed e-learning experiences for desktop and laptop computers. Many of us never really thought about e-learning as a "web product," although it's mostly delivered through a web browser. When the iPhone was introduced, Steve Jobs's keynote presented the mobile version of Apple's Safari browser displaying the *New York Times*. It seemed like a game changer at the time: a true user experience on a small screen that mimicked the desktop, no longer a special mobile-only view. Back then, Safari on the iPhone was often an unpleasant experience. Websites still needed to be optimized for small-screen displays, and performance was spotty at best. However, that presentation opened the door to the idea that there was a world beyond the desktop for consuming information, and ultimately for learning. Since then, smartphones have become extensions of our very selves, and are often the first place we go for information. Knowing this, how long can you ignore mobile technology to help your organization achieve its learning goals?

Many learning leaders ask themselves, "Why should my team develop mobile learning? If a workforce is tethered to a desktop computer for day-to-day tasks, why should I worry about mobile learning?" When you pick up your mobile device, you have in your hand one of the most advanced real-time collective content and learning experience mechanisms ever created.

In today's world, almost all workers are mobile in some sense. Constant connectivity has sparked a revolution in processes and agility, as well as rapid change in how business is conducted. Just because workers are tied to one place for a set amount of time does not mean they aren't using a mobile device. In fact, an important realization about mobile is that "being mobile" isn't just for workers on the move. Most mobile device usage occurs when people are in one location. Studies show that people always have their mobile devices, and often access devices in multimode usage; some tasks may be done on their laptop, while they pause and pick up their phone to complete a different task. People move back and forth between devices based on context, task, and ease of access.

Furthermore, people have integrated mobile devices into every aspect of their lives, including work, and are heavily reliant upon them for communicating, connecting, finding information, and sharing. Mobile in the workplace is really all about productivity, and its primary change to productivity is in shifting worker focus from efficiency to effectiveness. Would Amazon's two-hour delivery service be possible without mobile technology? From customer browsing and ordering to fulfillment and delivery, workers use mobile devices to ensure the two-hour customer promise is met. Workers use mobile devices to look up information at the moment of need as well as for real-time sharing of information.

Determining if Mobile Is Right for Your Organization

Mobile, however, poses major challenges for learning organizations. It's a technology that received mass adoption by the workforce, and now, in many ways, the workplace is playing catch-up. As entire industries adjust to the rise of mobile in the workplace, many job roles are affected. The debate about mobile for learning organizations is often polarized between those who see new channels for learning and those who see more distraction and less rigor around the formality of learning. The key is to determine if and when moving to mobile is right for your organization.

At its most basic, mobile is a confluence of devices, people, and connectivity with the ability to give a deep reach into the learner's context. However, not everything about mobile may be appropriate for your current needs. Like any of the factors discussed in this book, mobile can solve specific needs, but can also create a host of challenges:

- **Do you know how your workforce uses mobile now?** More than likely, they already use mobile devices. They may search with Google, use your consumer app (if you have one) to research information, collaborate with friends and colleagues using social apps, and use productivity apps to accomplish tasks. Do you have policies that allow access to mobile devices while working? Is it culturally acceptable to access learning resources while working? Is there Wi-Fi available and is it robust enough?
- **Is your team prepared to move from lengthy content creation efforts to curating and maintaining a living, breathing ecosystem of relevant, up-to-date content?** Mobile provides meaningful in-the-moment access for information and support. As a stand-alone learning tool, mobile is powerful, but when combined

with the cloud and the Internet of Everything, including wearables, it can open entirely new avenues for experiential learning.

- **Does your team have the skill sets mobile requires?** Resource shifting, upskilling, and integrating new systems takes considerable time and resources. Buying or building a platform, integrating it into existing infrastructure, and rolling it out require significant allotment of both budget and expertise across a wide range of capabilities. Ensuring you have the skills on your team to leverage mobile technology appropriately for learning is also vital. Designing and developing learning experiences for mobile requires a deep combination of both design and technical acumen.
- **Do you provide the device or does your workforce bring their own?** If you provide devices, then you need to manage and administer them. You need to consider the implications of "bring your own device" (BYOD) if your learning content requires authentication through the company network.
- **Does your company view learning as a formal event?** If your offerings are primarily classroom, how might mobile help? Formal classroom training will always have a place in learning. Some topics, such as leadership and sales training, are perfect for the instructor-led setting. In some situations, mobile might diminish the overall learning experience.
- **Can you gain support for the necessary technology shift to support mobile learning?** Many learning organizations rely on enterprise IT for technology support. In those situations, your tech support may come to you without substantial learning expertise, especially when it comes to implementing mobile learning.

Although there are challenges to integrating any new technology into your existing infrastructure, one reason you're reading this book is because you feel the need to understand how these factors affect your learning organization. An important element to consider is the simple fact that your customers are most likely using mobile devices to learn more about your business and interact with your company. In fact, customers are researching online and, with their mobile devices to help them, making buying decisions before they even entertain your product or service. Even if your workforce is tethered and not mobile in the classic sense, mobile learning can provide support for key business drivers, such as:

- **Offering faster training**. If you have a need for frequent training to support new products or services, mobile learning can reach your employees with product knowledge and selling skills. Consider offering all or parts of your mobile learning experience to customers as well.
- **Reducing training time**. Reducing time employees spend on nonrelevant training is critical, because now they can self-select the support they need for what they do, and then have more time to devote to work or customer service.
- **Reinforcing formal training**. Most training is intended to increase performance. Mobile enables learning organizations to increase performance by providing information when it's needed, rather than fire-hosing employees with information too early (or too late). By reinforcing prior learning, mobile can have a big impact on performance.

Mobile Technology's Evolution

Mobile technology is evolving separately from how people interact with mobile devices. In many ways, the mobile industry is still in the formative stage. For example, the voice-activated technology Siri turns the iPhone into a personal assistant that can complete tasks on command. Some Android devices sense your eyes on the device. If so, the screen won't dim, assuming you are reading. Although you are not tapping the screen, the device tracks your gaze, which could manifest still other behaviors, such as changing notifications.

Knowing how people use and behave with mobile devices is important. First and foremost, the devices that succeed offer the ability to keep in touch with a network of friends, family, and colleagues in a reliable, easy-to-use manner and encourage productivity. The technologies and trends detailed in the next sections are in their infancy now, but over the next decade will mature based on mobile technology evolution.

Mobile App Types

Currently, most developers choose from three types of apps: web, native, or hybrid. Web apps reside on a server and are coded once for multiple-operating systems. Native apps must be installed on the device and are written specifically for an operating system. Hybrid apps have components written in native languages for specific devices and are downloaded and installed to the device, but the content comes from the web. Developers are now blurring the line between the types of apps so that users are not aware when they link from one to another, or are not even required to download and install the app before seeing its content. Studies show that it's difficult to convince users to download apps, which means developers need to find new ways for people to discover content.

Mobile Payment Systems

Increasingly, smartphones are replacing wallets for everyday usage, such as payments. Amazon, Google, Apple, Chase, and others have all launched competing payment platforms. It's still very early in the adoption of mobile payments mainly due to financial regulation, merchant engagement, and technology integration. This is the one area that's evolving slowly, considering smartphones have been with us for almost a decade now. Just like commerce on the Internet took time to gain mass adoption, mobile payments will as well. However, in the future mobile payments will be a primary commerce mechanism.

Consumer-Led Mobile Health Monitoring, Diagnosis, and Wellness

Health monitoring is one of the most valuable applications for mobile because users always have their devices with them, and most smartphones have sensors that lend themselves to tracking activity. Viewing health data is becoming more and more popular as developers create apps that gather specific activity data. As the technology evolves, devices will do more than just track activity; they will prompt users based on contextual monitoring and behavior patterns, such as "Hi Michelle, don't forget to take your medicine; it's 3:00."

Corporate Adoption of Mobile Devices for Workforce Productivity

Many companies now see the value in providing devices for employees for tasks and to collaborate with colleagues during the workday. As the devices themselves become more commoditized, many more companies will adopt mobile devices for more employees. Some CIOs are also integrating BYOD policies that allow employees to access corporate servers and services with their own devices. Companies will mix the need to protect intellectual property with the need to ensure that employees are as productive as possible. Many employees now invest in devices and do not want to be required to use a less powerful or less feature-rich device than their own.

Sub-$50 Smartphones

As mobile technology evolves and becomes more ubiquitous, the cost of ownership is falling. In many areas, providers already offer feature-rich smartphones for under $100, and soon disposable smartphones will break the $50 price barrier. Cloud technology enables people to move from one device to another to continue their experiences. Eventually, people will have a preferred personal device, but also move between multiple devices to achieve work tasks. Smartphones will eventually be like distributing pens and paper—a cost, but not a significant one.

Producing Mobile Learning Content

RIM, the maker of Blackberry devices, made a strategic decision to ignore the burgeoning BYOD movement, and the iPhone became the desired device in the workplace. The learning organization needs to proactively move in the right direction, aligning with the business. However, adopting mobile learning requires a good understanding of the structural changes required for your team to effectively implement a mobile strategy.

Designing learning with a mobile-first mentality is becoming the standard for many learning organizations, providing more opportunities for multichannel access to your audiences. The basic philosophy of mobile first design is not new. In the late 1970s, Standard Generalized Markup Language (SGML), was created to provide descriptions for a document's structure and other attributes. Building on that is the idea to separate content from its display. By adding metadata (descriptions of the content) to the content itself, the content becomes portable and can be displayed on different devices and not confined to one style of formatting.

Most learning organizations will need to support both desktop and mobile, structuring delivery options to support multiple channels but minimizing redundancy in distributing content across those channels. Beware of setting up a system in which a mobile team is responsible for just the mobile content and a desktop team is responsible for just desktop e-learning content. In a perfect world, content is agnostic and can be easily optimized for necessary delivery channels, just like what the creators of SGML had in mind.

It's a waste of time and resources to develop one-off content experiences designed for specific delivery formats. Many content development processes and structures are still focused on delivering to nonmobile devices, and many learning teams start with thinking about delivering through the traditional web first and then converting that into a mobile experience. We have almost a generation of history and work flows to update, but it's important to evolve your work flow away from what your traditional systems dictate and move toward what content creators need when supporting mobile audiences.

Apply the same principles and techniques to your workflow design as you do to your content design. For the next several years, mobile usage will focus on the increasingly parallel world of technology and multiple devices. Even if you're not yet delivering on mobile, embracing a scalable content management and distribution system that supports mobile insures you against redesigning or redeveloping content across multiple channels. A cohesive, platform-agnostic content strategy is critical. Each of the technologies discussed in this book is directly affected by content, and vice versa, but none more important than mobile. Think about these questions as you formulate how to produce mobile content:

- Do you need to support various devices (smartphone, tablet, desktop, or wearables) across your workforce?
- Do you want to support single-source publishing?
- Do you have the necessary capabilities on your staff to design and develop for mobile?
- How will you support and maintain content?

One of the biggest challenges with adopting and integrating new technology is determining the level of interoperability with your existing infrastructure. Considering how quickly technology options evolve, it can be daunting to finally land on a solution only to learn there is something around the corner that may provide new, enhanced capabilities. As a learning organization, you want to provide a seamless experience for your learners, which means integrating with existing systems and devices. With mobile, it's even more daunting. As mobile evolves, it's fragmenting. Discrepancies among devices, operating systems, and online access make it difficult to develop experiences that are consistent across channels.

If your enterprise already has a mobile strategy that includes development of mobile experiences (a consumer-facing app or a mobile website), consider partnering with that department to leverage the technology and learn from their implementation. You may find the enterprise favors a specific mobile development platform, and there may be licenses to grant you access. By coordinating your mobile content strategy across the enterprise, you ensure deeper interoperability to curate existing content into your learning experiences. It's best to align to the enterprise strategy, if possible, but if you do have to go it alone as a learning department, consider the following approaches.

Don't Design for Specific Hardware Devices

Unless you're certain the enterprise is locked in on one device, it's important to remain device agnostic. If your content experiences can bend to any device, you'll ensure compatibility with the enterprise regardless of how they evolve. This is especially important if your enterprise allows BYOD for the workforce. Even if the enterprise has landed on one device for everyone, the lifespan of a device is probably shorter than the life of your content.

Focus More on Operating System Interoperability

Do your audiences use Android, iOS, Windows, or a combination of the three? You want your content to first work on the operating systems you are required to support. Your team should devise a "develop once, deliver anywhere" distribution method. How sure are you that the enterprise won't eventually move away from one operating system to another?

Create a Centralized Content Strategy

Try your best to remove your content from its presentation. For years, learning organizations have struggled with the idea of content management systems and how to most effectively acquire, design, develop, store, and deliver content. They usually have ended up combining content with formatting to create unique deliverables. For multichannel publishing, your first strategy should be to determine how to reduce redundancy and to have a scalable platform that appropriately supports multichannel publishing.

Understand That Mobile Learning and Performance Support Are Not Necessarily the Same

Mobile learning can be an experience designed to teach something. Mobile performance support can be viewed as helping the worker to perform a task at a moment of need. Mobile learning may be more of a "lean and learn" experience, whereas mobile performance support is task specific and needs to be accessed in the work flow.

How to Use Mobile Learning Content

Mobile can change the learning content you deliver. Here are some design affordances specific to mobile learning:

- **Device cameras for scanning and imaging**–You can use the hardware to make access to information easier and more natural. If your workforce moves around products during the day, scanning barcodes to acquire information may be quicker than entering text into a search query.
- **Geolocation for finding products, fellow employees, and customers**–Mobile devices have built-in sensors that provide location awareness, device orientation, and navigation. You can access that feature to help with wayfinding and ease of use in locating.
- **Internal apps for measuring and calculating**–You can access or point to apps to extend or augment learning: the calculator app, measuring apps if employees have to help customers decide how much of a product they need, or apps that provide product comparisons, ratings, or reviews.
- **Internet connectivity for search and collaboration**–Most devices have some type of connection. Constant connectivity means you can encourage employees to search or collaborate with others regardless of location.
- **Media viewer or playback for content**–Almost any type of media can be viewed, such as video, audio, webpages, PDFs, and documents.
- **Messaging for collaborating**–Every device comes with built-in apps for messaging. Messaging apps are not only becoming more common for communicating, but also great for photo sharing, short-range communication and document sharing, link sharing, instant feedback, and group discussions.

- **Microphone and audio recording for capturing ideas and suggestions to share**—Mobile devices are not only one-way information portals. Employees can provide input using the microphone, and video recording capabilities are included in almost every device. Think about the power of your employees providing their own training videos for their peers.
- **Notifications and alerts for time-sensitive information**—The ability to distribute near real-time information in an instant can help speed acquisition of important knowledge. If a product or service change occurs, content can be updated quickly and employees notified in multiple ways through devices.
- **Portability and mobility for moving around and connecting to printers and terminals**—This has become common in retail stores such as Apple, where you don't even see cash registers. Employees move around the store assisting customers and taking purchases with mobile devices. If devices are used for tasks, you can also let them be used for learning.

Determining the Effectiveness of Mobile Learning

When it comes to developing a successful mobile strategy, learning organizations are often faced with difficult considerations around the best way to measure effectiveness. The process of creating mobile learning and investing significant resources into it requires that you determine whether it provides a demonstrable return on investment.

With mobile learning directly affecting the performance and behavior of learners, it's easier in most instances to calculate the financial impact of mobile learning than it may be to calculate the learning, mastery, and retention of the information. Mobile learning tends to affect the learner at the higher levels of learning evaluation (Behavior and Results) compared with traditional training approaches (Reaction and Learning) (Kirkpatrick and Kirkpatrick 2006).

The first measure of success of mobile learning comes from changing how your learners consume learning experiences. By reducing the cost of learning development, reducing the seat-time commitment to longer learning, and providing more efficient transmission modes than traditional learning, the cost savings can be significant. Every business is different, and every worker context is unique. You first need to measure specific initiatives to get a true ROI. Next, look at efficiency over effectiveness in the short term to ensure your design and development strategy is correct for your environment. You also need to determine how to measure program benefits over program costs. In some situations, mobile learning may solve a broader problem with worker knowledge, performance, or morale you didn't know existed. To help determine the ROI of mobile learning, think about these tangible benefits:

- less need for formal learning (can have a labor-hour impact for nonsalaried workers)
- faster new worker readiness (on the job quicker, leading to improvement in productivity)
- improved customer satisfaction (better customer support or service, increased sales, more upselling).

When beginning to assess the efficacy of your overall mobile learning strategy, include the following considerations:

- **Identify mobile learning benefits.** These would include the ones mentioned (and the business objectives they support). Base them on your organization's specific context (people, processes, prerogatives).
- **Estimate usage of your mobile services.** Change management is necessary if introducing mobile devices is new to the company. If it's not new, estimate how usage of existing devices would change.
- **Calculate the total cost of ownership.** Devices are necessary. If you need new devices, you must estimate the cost of providing them, maintaining them, and supporting them. Even in BYOD settings, there's a cost to integrate and manage. Secondly, platform costs can be significant. Do you implement a vendor-driven software

as a service model for content development or an internal content management system, or do you augment internal resources? In many circumstances, you'll find "buy versus build" to be about the same cost overall. Given that, you need to determine if your internal resources can effectively support mobile learning from both the content and technical perspectives. For a first iteration, or to prove the concept, it may be worthwhile to partner with an external vendor.

- **Build a model to calculate the return over a period of time.** Using a proof of concept, or prototype, as a model can help determine whether mobile learning can work for the company. A "build small and iterate" style will uncover systemic barriers you may encounter. It's always best to start small and phase a rollout. Choose a program or project that provides the best representation of audience considerations and development strategies.

Ask the Expert: Tom King on Why Mobile Learning Matters

Even though workers have overwhelmingly adopted it, mobile learning is still in its infancy in the enterprise. What does mobile mean for learning in the near future, and what are the short-term considerations you need to think about? Tom King is a mobile learning expert who has worked in both learning and technology for more than 15 years. He is a technical learning adviser specializing in the integration of standards and platforms and has consulted with both corporate and government agencies on how to implement learning technology.

It seems like mobile technology has dramatically changed over the past few years. What's your thinking on where we're at right now and where we're headed?

We have a lot to be grateful for in terms of mobile learning capabilities. Modern hardware is more capable and pervasive than ever. Multimedia is now truly multisensory. Though it may seem otherwise, we're actually in a new period of relative stability for platforms. Dominant platforms have emerged and powerful capabilities are commonly implemented; although things will clearly continue to evolve, massive ecosystem-wide disruption is unlikely for a while. We are not facing the sort of sea changes we experienced with the vendor disruption of Blackberry and Nokia or the platform transition from laptops to smartphones.

I'd like to think that the enterprise learning and the e-learning vendor worlds will take this opportunity to catch up with the dramatic changes. Solutions, tooling, and governance for mobile learning still seems ad hoc and underdeveloped compared to the mass market offerings. Startups and innovators have had the nimbleness to develop workarounds for mobile learning that often feel just like that—workarounds. Much of the industry relies on desktop "rapid e-learning" tools based on paradigms that emerged 15 years ago.

Personally, I will continue to chase the hardware refinements and improvements, but I'm most excited about new tooling and models for mobile learning. Where are the tools to capture, create, or edit learning experiences using your smartphone or tablet? Mobile is by far the most widely adopted Internet and computing device, yet we continue to consider them second-class tools for creating the learning experience. We develop on desktop and "target" mobile for publishing, and that inherently isolates us from the native thinking and native experience of mobile.

So mobile is a big deal. Why do you think it's important for a corporate learning organization?

To connect, engage, and integrate with learners' lives, it is important for organizations to embrace mobile. It's a bit similar to the paradox of designing and developing exclusively on desktop while delivering on mobile. People embrace and rely on smartphones. They are how individuals connect, and how people engage with the services and businesses they use. Smartphones are integrated into our lives. It would be shortsighted to not integrate mobile as a key component of learning and training strategy. As the PC was coming of age, Microsoft had a mission statement of "a computer on every desktop." Over time that transitioned to "a computer on every desktop and in every home." Now we have the smartphone, a networked computer available in every hand. E-learning has, and will, follow a similar trend line. Computer-based training started with a mainframe for a campus, then computers on every desk in the training room, until the Microsoft mission was realized and we called it "e-learning" for the computer on every desktop and in every home. Growth and adoption of mobile has exceeded the rate and quantity of everything that came before it—faster than radio, television, or computers. With pervasive cloud services, we're now seeing more people with more devices, and not just the device, but the information is accessible everywhere. Traditional desktop and laptop computers are now the minority of devices connected to the Internet; mobile phones, equipment, and the Internet of Things are the vast majority. Mobile is important if you want to reach your largest audience, with the technology that is most widely available, and that they choose to use most of the time.

How can learning leaders prepare their teams to be able to deliver mobile learning?

Three things come to mind regarding preparation: contradictions, empathy, and collaboration. First, it is important to realize that mobile is filled with contradictions. Mobile is both pervasive and fleeting, it is engrossing and distracting. It is cheap and it is costly. It is completely easy to do and seemingly impossible to do completely right. An important part of preparation is accepting the contradictions and making trade-offs. The trade-offs must be carefully weighed, executed decisively, and then scheduled for re-evaluation. It may seem like a great idea to make training on critical items entirely mobile, to get that wide reach and exploit the pervasiveness. But does that suit the topic? Would it be wiser to engage mobile as part of a blend or to break it up into smaller pieces? We spend a lot of time doing things without our smartphone, yet we only spend a short time doing things with our phone. The apps are small and focused; even if we have ostensibly one thing to do, we still hop from app to app. We get the reminder, check the flight, get the file from the cloud drive, send the email, call the Uber, and send the text. Sometimes we even answer a call. This behavior becomes deeply ingrained and reinforced. Initial acquisition of deeper concepts or complex skills isn't suited to such fractured, interrupt-driven experiences. Yet, those same stolen moments of attention could be exploited by e-learning used as gamification for removing some basic skill acquisition drudgery, drill-and-practice rehearsal for fluency, or spaced learning of small bits over longer intervals for increased retention. Evaluate and embrace the contradictions. Play the role or designate a devil's advocate, and more knowingly, make the better trade-offs.

Second, it is important to have empathy. The learner with a mobile device is also a doer. Mobile is a context, not a platform. Engage in your own ethnographic research and immerse yourself in that learner-doer's world. Use their type of device, in their type of environment, with their interruptions, demands, distractions, and deliverables. Then you can bring your understanding of organizational and training goals into their context. Mobile learning must work in context. Consider the field technician who might have spotty network service, cold or gloved hands, loud equipment nearby, or a device screen washed-out with direct sunlight. Consider the customer service representative who might need to break away from the training at a moment's notice, or may have a cancellation that suddenly leaves them with an hour of free time and only access to some "three-minute trainers."

Third, collaboration. Empathy resurfaces in regard to collaboration. In a larger organization you must collaborate with multiple teams to succeed with mobile learning. You will need to have empathy for line-of-business management, mobile device management, IT personnel, accounting, legal, marketing and branding, information security, and others. Oftentimes, they will need to educate you, and often you will need to research issues and educate them (or even selectively push back on some of their assumptions or assertions). In larger organizations, the IT department will often have significant investment in mobile device management and IT resource to support field teams who have mobile devices to support operations. However, training rarely has a seat at the table when decisions are made for such mobile deployments. Be ready to collaborate with the line-of-business managers who drive the initial operational justification and deployment. Work with the IT team responsible for deployment. Check with colleagues in your organization and peers in analogous operations. You may be surprised to find out about issues with accounting for international data plans and app purchases, global trade concerns about encryption or data security, or even shipping concerns. I know of one organization that was ready to deploy a tablet-centric "e-classroom in a crate" only to discover at the last minute that a box of 24 tablets with lithium-ion batteries is considered hazardous materials and could not be sent on international flights. Remember to collaborate with your designers, developers, content publishers, and LMS vendors, too. They may have answers or they may need to be educated—don't depend on assumptions, count on collaborations.

Is there anything else you can think of that you've come across that learning leaders are wrestling with when it comes to mobile?

Yes. Many leaders or their colleagues assume that once mobile is in place, you're done. However, all those collaborations are moving pieces. A mobile strategy is a vector, not a location. You need to revisit things and should create your own monitors, metrics, and "seasons" to manage a mobile strategy because it is an ongoing process, not an event or a reaction. Consider how you might align the entirety or portions of your mobile strategy based on indicators and external factors such as annual or quarterly company goals and initiatives, IT infrastructure updates and schedules, and supplier schedules. (The major fall releases of iOS and Android can wreak havoc on slow-to-update apps.) Consider strategies to use off-the-shelf apps like checklists or PDF viewers for documenting on-the-job training observations—and be ready for when those apps might need to be replaced. In short, be informed, be agile, and own your schedule as much as you can.

Case Study: Mobile Learning in Retail

In just the last several years, mobile technology has transformed the retail world. Customers use mobile devices before, during, and after transacting with retailers to ensure their product and service needs are met. Customers are becoming accustomed to receiving hyper-individualized attention as they make buying decisions, and they're using mobile to stay informed and aware. In many ways, retail customers are entering the buying journey with more information than the associates that serve them have.

Recently a large, U.S. big-box retailer embarked on a mobile learning pilot designed to help improve customer service by bringing knowledge for both their associates and customers into the aisles and out of the back rooms. The company has a large mix of part-time and full-time workers and a highly complex retail environment with a dedicated focus on customer service. The bulk of their current training is provided at the start of employment and consists of approximately 30 to 40 hours of selling skills and product knowledge training.

The retailer piloted a mobile performance support system designed to provide associates with product knowledge in the aisle to help customers with their product and project needs. The app is accessible from corporate devices or the employee's personal device and provides quick access to relevant information about top-selling products and popular projects.

In this context, mobile learning adds meaningful information at the moment of need instead of providing traditional training on products outside of the employee experiencing the product in the aisle. Before the mobile learning pilot, observations showed that employees learned by using their own devices to search product reviews and ratings, check inventory, and locate products in other stores.

A key reason this retailer is piloting placing a device in the hand of every sales associate is because it recognizes that learning happens anywhere, anytime, and most workers learn how to do their jobs while on the job. The ability to move about while remaining connected to information that helps the employee perform is the essence of learning embedded in the work stream. The employee equipped with a mobile device is similar to the employee equipped with a pair of gloves and a measuring tape. It's an essential utility to extend their knowledge so they can continue working with customers, instead of going off the floor and sitting in front of a computer for e-learning. Furthermore, customer research shows that the number-one customer desire in a retail shopping motion is to receive quick and knowledgeable help from an employee.

Initially the business wanted the pilot to answer these three questions:

1. Can mobile learning be used for training in the aisle?
2. Will it increase associate speed-to-readiness for the workforce?
3. Will it increase new associate competence and confidence to serve customers?

The retailer felt if the pilot showed positive data in each of these three areas, there would be justification to move forward with integrating mobile learning into its overall learning strategy.

App Development

The company decided to first construct a proof of concept with a small amount of content and limited functionality. The development team applied design thinking principles by going into the store to design a series of conceptual mock-ups. This enabled the team to get immediate feedback from the target audience. By ideating a proof of concept with the learner involved, the team quickly established a user-centric design pattern, shaving a considerable amount of time from the typical design process. For example, developers considered search functionality to be a core component of the user experience. They could quickly discern that most learners would not use it. Instead, they wanted more guidance through a menu system. (Consider the resource savings of not engineering functionality that won't be used.) Additionally, approaching the content development in design thinking mode, the prototype tested not only the effectiveness of the content itself but also how employees would access and use it. Usability and content together form the mobile experience, and the prototype was designed to test both the content and how employees might use the app.

From Prototype to Pilot

Once the prototype was constructed, it was delivered to 33 stores. The primary use case for the prototype was to observe employees using it in the context of their jobs, gauge reaction, and analyze behavior patterns. The team also focused on customer reaction to employees using the app while working. A key component in rolling out a mobile learning experience such as this is to identify how its usage in the workplace affects not only employee behavior but also customer behavior. It was also important to receive store leader feedback to discover whether the use of devices in the aisle affected customer service.

The app was well received during the prototype, which ran in the stores for 90 days. However, access and usage of the app over time suffered because of device limitations: Notifications were turned off and the app icon was not easily accessible.

Pilot

The team took the feedback from the prototype phase and iterated the design. The content was optimized to be more succinct and provide clearer guidance on helping the customer make purchasing decisions. The team also integrated game mechanics to provide a more interactive and engaging learning experience. With the addition of guided scavenger hunts and rapid-fire knowledge checks, the employee could leverage the app to not only help customers but also become more fluent in the work environment. The game mechanics assessed the employee's product knowledge, and using reward mechanisms such as points, badges, and leaderboards provided a competitive and challenging learning opportunity.

The pilot was conducted over a 90-day period in 78 of the retailer's stores. A control group was established to benchmark against the pilot group to help the team determine a return on learning effectiveness. Metrics such as app dwell time (the time spent in the app itself), motivation to complete game activities, and proficiency based on assessment responses were analyzed to determine the pilot's success.

Evaluation

For a large retailer, it was important to prototype and pilot a learning experience that would potentially transform how it delivered learning before launching a large-scale rollout. Iterating through the design and development experience by connecting closely with the target audience allowed the team to reduce their reliance on assumptions and deliver to actual audience needs. Another key aspect of determining whether mobile would work was to recognize the change management necessary to facilitate learning in the aisle. From the employee's perspective, the leadership perspective, and the customer perspective, it was important to fully understand how (and if) the use of devices for learning in the work stream would have a detrimental effect on the business. The team found that quite to the contrary, customers, leaders, and employees welcomed the experience into the environment because at its core, it was designed to assist.

Key Takeaways

Of the five factors, mobile is the one that can demonstrate the most tangible ROI for a learning organization in the short term. We all know that more people will be typing on glass rather than keyboards in the future. Even if your workforce isn't mobile in the sense of moving around, they constantly use mobile devices to acquire information and learn. Having a learning strategy that incorporates mobile prepares you to accommodate your learners across the channels they feel most comfortable with and to which they have access. You should never incorporate emerging technology just because you can. You should also remain skeptical of emerging trends. However, mobile technology has completely reshaped business and provides continuing value as to how people get work done. Consider these guiding principles when thinking about making the move to mobile learning:

- Place learning opportunities where it matters most for the workforce: at the point of need, where it's more relevant and engaging.
- Accelerate the creation of learning content and increase the speed of access to keep your learners informed with up-to-date information.
- Stop the information fire hose and provide just the right amount at the right time.
- Untether your workforce from "dedicated training computers" and make information available anywhere, any time they need it.

- Leverage geolocation capabilities, internal sensors, text messaging, supportive notifications, collaborative learning, and smaller chunks of content inherent to mobile to redefine learning experiences.
- Move from formal learning to self-directed learning where the employee has more control.

4

Big Data and Analytics

"The temptation to form premature theories based upon insufficient data is the bane of our profession."
—Arthur Conan Doyle

JANET LEE BEGAN HER FOUR-HOUR SHIFT AS she always did, configuring her handheld device for the job ahead: tracking the myriad helicopter parts that would flow her way so she could get the right parts to the right area for repair. Janet works at the U.S. Navy's helicopter repair facility in Corpus Christie, Texas. A typical helicopter has thousands of parts and the repair facility attaches a radio frequency identification (RFID) tag to each part as it's disassembled for repair. The RFID tag identifies the part in real time through the repair process and helps technicians correctly reassemble the helicopter with the same parts after the repair is complete. Whenever Janet needs information on a specific part, all she has to do is use her handheld device to scan the tag and view the part's current status and maintenance history.

An RFID system consists of three components: a tag, a read-write device, and a host system for data collection (often a computer). The tag itself usually has its own power source, which allows it to transmit information. RFID tags are not a new invention. When combined with more recent innovations in big data, the tags enable companies to more efficiently track their assets, maintenance, and repair processes, as well as control cost through more accurate inventory control.

By using technology such as RFID tags to streamline the repair process, reassembly is shortened and there are fewer lost or incompatible parts to replace. By combining data about the part's maintenance history, technicians such as Janet are able to make quick decisions about its viability. The tools created to enable workers like Janet to filter reams of data across the repair work flow has increased productivity and efficiency, resulting in more timely and accurate repairs. This process represents a good example of how leveraging data in the work stream can drive performance improvement and reduce the risk of failure.

There's nothing more powerful for business than the data it can gather, as long as those data can be appropriately managed, analyzed, and filtered to improve the bottom line. Janet can leverage technology to make timely decisions to positively affect productivity. The integration of data, technology, and processes drives efficiency gains across a broad spectrum of business functions, with the underlying support being the data.

The Big Data Revolution

Let's take a deeper look at what big data is. *Big data* refers to large data sets: data that are typically too large for the average software to gather, store, and analyze. Big data analytics is the process of analyzing big data sets to uncover business-relevant information. The concepts behind big data have been with us for quite some time. For years, companies have been gathering data and analyzing it; however, recent advances have enabled real-time data capture and analysis, which has led to much quicker processing of data, enabling deeper insight and leading to more immediate decision making. Basically, big data is driving the ability to work faster and smarter, giving companies a competitive advantage. The best way to look at it is to think about a specific business problem and determine where data can be used to formulate a solution.

In some ways, the term *big data* itself is a misnomer. The value derived from a large data set is not due to its sheer volume—its true value rests in the unstructured nature, or variety, of the data. By examining diverse sources in the data, patterns and behavior can be uncovered, which help shape a perspective about what the data represent.

The Value of Big Data

Before almost everything was smart and connected, data were collected through separate transactions and combined using surveys, research, and other external sources to derive actionable information. This was a somewhat decentralized manner of manually bringing data sources together to synthesize information. Now data can be gathered from the experience as it occurs, making the data a core component. For a learning solution, the content, the learner, and now the data combine to provide valuable information about the effectiveness of the experience. This is where the learning organization can augment an often-vague ROI measurement with a more insightful return on learning effectiveness (ROLE) measure.

The value of real-time data increases exponentially when it is linked with other historical data, such as job roles, certifications, curriculum completions, and promotions, to help you gain performance-related insight. Its real value is aligning career development to the learners over the life of their tenure, and if possible, previous experience. Knowing the past achievement of learners beyond just their course completions can help identify and develop the high-performing employees or the ones who have the potential to become high performing.

Collecting massive amounts of learner information also requires effective visualization mechanisms to extract actionable meaning from the data. The goal is to uncover trends, gain insight, and derive a deeper understanding of your audience through the data you collect. The transformation data bring is not in the data themselves, but in what you can accomplish with them. You can mine data to drive correlations; lower-performing employees receive reinforcement on basic processes and procedures, whereas average to higher-performing employees receive different activities or stretch exercises to help move to a higher level of proficiency. If they get stuck, or need reinforcement, the system can adapt.

Big Data in Learning Organizations

A common use of big consumer data includes recommendation engines created by Amazon, Google, Netflix, and others to assist in purchasing or viewing decisions. This playlist idea can extend to learning to serve personalized content experiences based on an individual's capability. This is not new, but with data the learning organization can be more assured that the recommendations delivered are appropriately tuned to learners' current abilities and can facilitate their progression.

Today's learners want to be treated like individuals, which means developing learning experiences that adapt to their preferences, generate tailored results, and anticipate their future needs. Learners need to know what they *should* know, and you need to know what they *already* know.

Today, when assessing the competencies of individual employees, we tend to look at their past, including their resume, diplomas and degrees earned, and training they have completed. Now, with big data and analytics, you can access more real-time data to gain a more comprehensive view of their capabilities as they work. Correlating these data with their past data can provide insight into their future potential.

The data collected across both the work and learning context can be combined to show real performance metrics. Alfred Remmits, an expert on adaptive learning, implores learning organizations to seek causation by asking the following:

- How often has the worker performed specific tasks in the last six months, the last 12 months, or the last five years?
- What feedback has the worker received from customers or stakeholders on the performance of those tasks?
- What feedback has the worker received from managers on the performance of those tasks?
- What is the feedback from peers?
- How do you rank the feedback? Should the opinion of a customer or stakeholder be weighted higher than the peer feedback?

- What other data can be obtained about the worker's performance that paints the complete picture of the worker's current capabilities and potential (team member satisfaction rates, customer satisfaction rates, revenue figures, profitability figures, and so forth)?

By bringing these data together, it's possible to create a link between the real work performance of individuals and their learning needs. There are multiple applications of how data can be leveraged in real time. One often-missed component in learning design is consideration of the learner's environment. If the learning experience is not event based in a formal learning setting and instead occurs in the work stream, you can build associations with the environment itself. For example, in an industrial environment, linking learning content to help a learner more quickly grasp how to use hardware while using it can go much further in increasing performance than creating a training course or video that's viewed outside the context performance.

Knowing if and how the environment itself can be used for learning can also help you design deeper associations for learners. If learners can use their devices to view information about a product, consider providing a learning opportunity using that medium to help them gain deeper insight about the product by encouraging them to engage in a hands-on activity with it. Leveraging data in the moment is akin to performance support on steroids.

Evidence of learning efficacy through data analysis is becoming the single-most powerful instrument the learning organization can use to demonstrate its value to both the business and the learner. Additionally, big data can help you gather deeper insight into the overall learning population. The data will provide information about all your learners, as well as individual learners throughout the same experience. When learners consistently answer the same questions incorrectly, they immediately receive individualized feedback based on their input and the input of other learners over time. This helps create personalized learning experiences and enables instructional designers to make rapid modifications to the course content.

When integrating data into the learning strategy, consider the downside to large data sets. It's possible to build correlation without fully understanding causation. In other words, making assumptions about what the data may tell you is risky until you investigate. You can get answers from assessments, but characteristics around how and why the learner answers in a specific way are harder to uncover. Decide what data you should collect. For example, is it important to know how long it took a learner to answer the question? Consider delivery methods. Some learners may be mobile while others are not. We know user behavior on a mobile device is different than on a desktop computer. Perhaps you see a drop-off in usage of a learning experience at a certain time of day in a specific region on a continuous basis. It's possible you will discover that the learning solution is having no impact on performance, allowing you to more quickly intervene. Observing usage directly may show you other causes that data don't show. The drop-off could be happening because of other factors or conditions, or managerial solutions. Big data provides more transparency much sooner into the learning experience, but you must have a good understanding of what the data mean, as well as what they do not.

The Shift to Data-Driven Learning Design

The collection and analysis of big data brings new challenges for the modern learning organization (Marr 2016). The sheer size and speed of the data collected can strain the sturdiest network infrastructure and create a flood of data from which to analyze and report. Many learning organizations are creating job roles solely dedicated to the collection and analysis of big data. This role is typically responsible for gathering data and analyzing it to find correlations to key performance indicators (KPIs). By seeking deeper insight on the success of learning solutions, the process of analyzing the data from a learning perspective (referred to as learning analytics) can provide valuable information on what components of the learning experience drive productivity and engagement. Analytics offer a great opportunity to view real-time impact of the learning.

Ask what data you currently gather to inform you about the effectiveness of the learning you deliver. You may gather data through a centralized learning management system (LMS) and store course starts, completions, and assessment responses and scores. These data may be used to confirm completion reporting based on compliance or an confirm enrollment requirement, including certification or accreditation. These types of summative data are important in the process of determining the value of your learning experiences. However, it doesn't give you a deep insight into what occurs during the actual learning experience. Learning analytics help you derive information about what learners are doing during the learning experience. These formative data are distinctively different from assessment; they provide insight into learner behavior that uncovers patterns, such as where learners are struggling, or where they are not deriving value because they spend little time in a specific area. These data inform you about the value of the content. In this context, learning analytics lead to actionable insight, enabling instructional designers to be more responsive and gain a deeper understanding of learner context that can inform their learning design.

Analyzing granular data also leads to deeper insight based on several data points, demographics, job roles, geography, and schedules. The ability to discover under performing learners with correlation to performance issues can also inform how you deliver business value. For example, by viewing data based on a specific job role in a specific geography, which also correlates to underperforming business, you can connect the dots to how a learning solution may or may not provide a distinct return to that business problem. When one region is underperforming in sales, learning analytics data can expose behavior patterns that inform you on what parts of the content have the most value. This group of underperformers may have different needs than other groups. Your learning strategy may need to be modified to support this specific group.

This is a positive development for learning organizations because in most instances, the design and development of learning experiences begin with some type of analysis that outlines the problem. This leads to assumptions, that could be wrong without the luxury of real-time data. You can say goodbye to assumptions from the front-end analysis that are often hard to confirm. Granted, this is one dimension of effectiveness, but it can offer more insight than you could previously gain before real-time data collection. The promise is not in the quantity of the data you collect; it's knowing that now you have the information to create something meaningful.

The Data-Oriented Learning Organization

Analytics for learning is certainly becoming more of a mainstream idea, but it has yet to become a mainstream practice (Kiron et al. 2014). There are several opportunities to gain from data analytics for learning organizations, but there are also challenges. At first glance, it seems natural to assume there's significant business value to gain from gathering deep insight into the usage of the learning experiences you deliver. As the industry's ability to collect granular user data and parse through it to uncover patterns matures, the implications are becoming clearer. Identifying user behavior trends enables you to more acutely determine what types of content resonate, what doesn't add value, and how influential the content is in moving the performance needle. As you make use of analytics, you'll discover much about the effectiveness of the learning you're delivering, your value add, and the depth of your audiences' expectations.

Make sure you have the correct assumptions. Define what drives your workplace performance. Consider if you're relying on assumptions, or whether you have any type of exacting data to inform you. Research on the "service profit chain" identified four employee performance measures that drive service, resulting in increased customer satisfaction (Heskett et al. 2008):

1. internal service quality
2. service capability
3. employee satisfaction
4. employee loyalty.

Service capability and employee satisfaction have direct links to training. It's increasingly important to provide evidence that training affects at least two performance measures: employees' service capability and their satisfaction. It's no longer tenable to deliver training without a plan to measure its ROLE, which drives ROI. To verify that the program begins on the right track, consider these steps to map outcomes to specific performance measures:

1. Link every training program directly to a KPI for the business.
2. Using data analytics, measure how the training program correlates to the KPIs.
3. Build a map or diagram to show the connections from the training programs to the KPIs.
4. Identify where multiple programs intersect or map to the same KPI. Evaluate the programs to uncover any redundancies between them.
5. Evaluate the findings to determine if there are missing linkages, overlaps, and unexpected connections.

The findings from this exercise may help relieve you of relying on assumptions or, at the very least, provide insight to either affirm or deny the veracity of those assumptions. Additionally, striking a balance between assumptions and data will lead you to a more data-centric mindset when it comes to determining the value of your organization's contribution to performance measures. You will always make assumptions about performance measures—many of them based on valuable and experienced judgment of the business, the audiences, and the cultural specificities inherent to the environment in which the performance occurs. Data establish a firm foundation on which to derive a solid understanding of the true drivers and real gaps in performance. By applying a more rigorous approach to gathering and analyzing data, you will move away from accepting general hypotheses about performance that may not be accurate and develop more effective learning experiences going forward.

The Genesis of Analytics

Most technology adoption in large companies begins in the IT organization, and this is also the norm for data analytics. Over the last several years, many companies have been delaying large IT expenditures and pivoting to data analysis to inform them about appropriate investments in their IT infrastructure. In some respect, this movement has placed experience over data analysis in decision making. Where is the intersection between human analysis and data analysis, and what are the implications for relying almost completely on algorithms to make business decisions?

Data Analytics

As the ability to unlock the full value of data becomes a key source of competitive advantage for business, the management, governance, analysis, and security of those data is developing into a major new business function (Porter and Heppelmann 2015). Many learning organizations struggle with how to use their data. To be successful as a data-driven learning organization, you must align your plan with the company's overall strategy. Without a direct alignment between the learning strategy and the company strategy, the data will not provide the correlation you need to realize ROLE.

There is an investment required in technology as well as in team skill sets. Companies can discover powerful insights by identifying patterns in data sets produced over time. For example, information from disparate individual sensors, such as a car's engine temperature, throttle position, and fuel consumption, can reveal how performance correlates with the car's engineering specifications. Linking combinations of readings to the occurrence of problems can be useful, and even when the root cause of a problem is hard to deduce, those patterns can be acted upon. For example, data from sensors that measure heat and vibration can predict an impending bearing failure days or weeks in advance. Capturing such insight is the domain of data analytics, which blends mathematics, computer science, and business analysis techniques.

Analytics in general is about making better decisions. You gather data to understand what happened, what is happening, and what needs to happen (hindsight, insight, and foresight). You leverage data analytics tools to extract meaning from the data. Those tools fall into three key categories: descriptive, predictive, and prescriptive.

Descriptive Analytics

Descriptive analytics looks at past performance to help you gain a good understanding of how something happened. Descriptive analytics lays the foundation for converting raw data into useful information.

Predictive Analytics

Predictive analytics turns data into valuable, actionable information. Predictive analytics enables advanced forecasting to anticipate future results.

Prescriptive Analytics

Prescriptive analytics is where real-time data kick in to make predictions and provide recommendations based on learner input. Prescriptive analytics anticipates not only what will happen and when, but also why. The key aspect of prescriptive analytics is to combine unstructured data such as text, images, audio, and video from multiple sources to provide the right content at just the right moment.

Using analytics across the learning organization can proactively help with retention, ongoing learning (success and failure areas), and recruiting. By exposing real issues in the workplace, you can gather information that will lead you to understanding what is happening and why and then enable you to act.

As you build an analytics strategy, data from a range of sources can offer a more informed view of the individual learner's needs. Multisource data can enable a customized experience that matches people with the capabilities their jobs demand, distinguish reasons that may be causing performance problems, and help learners identify the skills they need to develop to advance their careers. These three components will support your analytics strategy:

1. software tools that enable frequent and real-time measurement
2. ability to integrate data from disparate sources
3. ability to analyze data into actionable plans.

One value proposition underlying real-time analytics is the fact that you can do more with less. The data collected uncover what matters most to your audiences, which can further inform your learning strategy. More important, analyzing the right data in real time enables a sophisticated learning system that dynamically responds to input and can more readily adapt to each learner's needs and expectations.

With learning analytics, you meet the learners where they are, can anticipate where they will go next, and enable better interactions at every stage of the experience. Real-time analytics will move you beyond assumptions that at times challenge what occurs in the workplace. Use real-time analytics both to dispel wrong assumptions about the capabilities of the workforce and to predict workforce performance. Too often, learning organizations have been reactive to the performance challenges in the workforce. Because these disparate sources of data are available to help business solve problems, it makes sense that the learning organization leverage those data to inform the learning strategy. In some companies, it's the learning organization that best knows the culture of the company, how employees behave, their challenges, and where they succeed. This knowledge can help in analyzing the data to derive the questions that guide the behavior change.

When evaluating support tools, such as human capital management software, in creating an analytics strategy, you should understand the impact of the software in addressing specific business challenges. If you aren't connected to HR in any way, it's important to determine how you aggregate and integrate the disparate data sources into a construct that provides meaningful information. Most companies will have different systems for different processes, such as talent acquisition, performance management, learning and development, and compensation. Your challenge is to take advantage of data sources across all systems to get the view you need.

Ask the Expert: Jason Haag on Big Data and Learning

Jason Haag has more than 15 years of experience in education, training technology, and distributed learning systems. Jason specializes in strategic integration and implementation of learning technology for the Advanced Distributed Learning (ADL) Initiative, a U.S. government program that conducts research and development on distributed learning and coordinates related efforts broadly across public and private organizations.

Can you talk a little about your work at ADL and some of the initiatives you're focused on?

ADL collaborates with government, industry, and academia to promote international specifications and standards for distributed learning content, services, and systems. Our standards and specifications, such as the Sharable Content Object Reference Model (SCORM) and the Experience Application Programming Interface (xAPI), have had a huge impact and been adopted internationally, not just by the Department of Defense (DoD). Since ADL's inception, we have fostered the development, dissemination, and maintenance of guidelines, tools, methodologies, and policies for the use of distributed learning resource sharing across DoD, other federal agencies, and the private sector. We have also supported research and documentation of the capabilities, limitations, costs, benefits, and effectiveness of distributed learning. I'm involved with research, development, and implementation of learning technology. I've been mostly focused on SCORM, xAPI, instructional design, mobile learning, and semantic web efforts at ADL. I'm also the technical point of contact (TPOC) and manager for several contracts funded to address ADL's broad learning science and technology research objectives.

What's a good definition of big data from a learning business perspective?

Big data was historically referred to as a way to describe data sets that were so large in size that processing and understanding the data was significantly complicated. The original definition and focus was on the size and complexity of the data and how to best access and manage it. For some organizations, dealing with hundreds of gigabytes of data was significant. For others, hundreds of gigabytes of data were manageable, but dealing with hundreds of terabytes of data was daunting. In August 1999, one of the earliest organizations to use the term *big data* was the Association for Computing Machinery. In Communications of the ACM, mathematicians and computer scientists revealed that the true purpose of working with data is insight, not raw numbers. Today, from a learning business perspective, *big data* doesn't necessarily refer to the size of a data set alone. Big data is an umbrella term that generically refers to the use of predictive analytics, user behavior data, or contextual data analytics that extract value from one or more data sets, regardless of size. Another way to think of this use of the term is that you can do "really big things" with the data. The buzz around big data has become much louder in recent years in the education and training world thanks to the big things you can do with xAPI. The xAPI is a technical specification and a means for collecting, reporting, and gaining analytical insights on big data associated with learning experiences.

How does big data affect the more traditional methods of learning effectiveness evaluation, such as the Kirkpatrick Model?

The Kirkpatrick Model is known as one of the most widely referenced yet often partially implemented evaluation frameworks for training. The only limitation of the Kirkpatrick Model is how it is applied in practice, and sometimes the cost associated with fully implementing it drives that practice. In many online training scenarios, organizations only require minimal evaluation at Levels 1 (Reaction) and 2 (Learning), and rarely evaluate or look for changes in Levels 3 (Behavior) or 4 (Results). Externally developed training courses commonly only measure up to Level 2 due to the cyclical nature by which the content is procured, developed, delivered, and forgotten, without budgeting or planning for deeper evaluation. As a result, quite often only the rote knowledge of the learner is assessed. By using the analytics and insights from big data, we can now more easily measure changes in behavior because of the training experience. We can also correlate big data from training experiences to business data and increased productivity output of employees. For example, imagine a sales training course that measures Levels 1 and 2, and a companion mobile performance support or job aid that can be used to provide additional support after the training is completed. Data is collected before and after the training, and additional data is collected to measure electronic-based behavior changes, as well as access to specific support materials that are designed to help improve sales skills. This repeated access and use of job support materials is critical data that can be correlated with other business data, such as employee sales performance or even applying important corporate policies or processes that influence sales results. Ultimately, you could understand more about the effectiveness of both the training and the process materials maintained outside of the training. Collecting big data in support of all four levels of the Kirkpatrick Model can yield useful insights about changes in behavior and performance outcomes related to the training, but it can also yield deeper insights into the relationship between the other levels and other business resources as well. Insights gathered from big data may lead to process improvement, increased productivity, and more confident decision making. All of these can further result in greater operational impact and employee efficiency, cost reduction, and reduced risk.

How do we begin to design learning systems that can provide insight into its effectiveness as the data is collected?

Assuming you have buy-in from executive leadership, learning systems that share data and use data with other systems or applications must all be designed and architected to support integration with external web services, APIs, and open data formats. The delivery systems alone shouldn't be evaluated for effectiveness, but effectiveness of the education, training, or performance support materials that are delivered should be evaluated. To design learning interfaces, activities, and content in a way that can provide insight into its effectiveness in real time as it is collected, consider dashboards designed in a way to provide immediate feedback loops to instructors or trainers. These types of interfaces could provide immediate insights into their effectiveness as the data is collected.

As the workforce becomes more mobile, how does big data help people adapt their skills at a faster pace to keep up with the demands of business?

Today we live in a world where collecting big data is not much of a challenge. The bigger challenge is deciding if specific types of data are more important than other types, and how to leverage the important data to improve skills and performance. Mobile apps are constantly capturing our usage behaviors, interests, and knowledge about topics and then using that data to make predictions, recommendations, or even to improve products. Why not use that data to improve skills or business performance? Big data results can be aggregated and expressed in a dashboard or even recommendation engines to help guide or inform those looking to adapt or improve their skills. This approach would likely need to tie into an ecosystem strategy. Big data alone is not going to be sufficient, but making that data actionable could be really powerful if you can direct the learner to take action to continually encourage professional development.

In your opinion, how do big data and analytics improve the learning team's capabilities?

Big data can provide an opportunity to conduct deeper levels of evaluation as discussed previously. The data obtained from evaluation can be used to improve the design quality of the learning content. The learning team's capabilities and overall value they provide are now increased to support human performance and, ultimately, the bottom line. If they are responsible or connected to productivity or any form of business results, then the importance of their roles should increase dramatically. For good reasons, the instructional design practices for SCORM e-learning content in the past were largely limited to the cognitive domain. In other words, SCORM was useful for recording knowledge-based learning outcomes but nothing else really. Big data now provides more opportunities than ever to also look more deeply at the connections between Bloom's three domains of learning: cognitive, affective, and psychomotor (1984). Learning teams now have opportunities to improve their capability to create learning experiences and activities that can provide significant performance value. This focus on improving performance and augmenting skills—rather than on just knowledge transfer—can result in generating learning objectives and activities that are more meaningful for determining common skill deficiencies.

Can you talk a little about what xAPI is, how it's different from SCORM, and its practical use cases?

The growth of the Internet forced an evolution of computer-based training in positive, but disruptive, ways. We said goodbye to delivering training content on compact discs and hello to embracing technologies such as HTTP, HTML, Flash, and JavaScript. In addition, learning standards such as SCORM provided new opportunities for tracking a learner's progress through interoperable, plug-and-play content. However, the data collected was locked inside an LMS and was generally limited to recording the progress of one user's interactions, scores, and completion of the content. This decade's disruptive technologies, such as mobile devices, augmented and virtual reality, social media, and big data, have stimulated a new market for ubiquitous learning and open data and paved the way for the creation of xAPI. In general, APIs define how disparate software components can openly communicate and exchange data. xAPI enables tracking of any electronic experience or activity and even some real-world activities. Unlike SCORM, xAPI is not limited to a browser for delivery and supports tracking activities on any platform or software system, including mobile devices, simulations, wearables, sensors, and more. And also unlike SCORM, xAPI provides open access to learning and performance analytics as it focuses on the data instead of specifying rules for content itself. xAPI can track several types of microbehaviors where data can be captured. These are some basic examples:

- reading an article or interacting with an e-book
- watching a training video, stopping and starting it
- training data from a simulation
- tracking performance in a mobile app
- chatting with a mentor
- monitoring physiological measures, such as heart-rate data
- engaging in microinteractions with e-learning content
- participating as a player in a multiplayer serious game
- recording quiz scores and answer history by question
- performing in a real-world operational context.

By design, xAPI enables new opportunities for tracking any type of experience in any environment with or without a web browser. In summary, xAPI enables the following in a more open and interoperable manner:

- Distributed content—Learning content can be hosted on a local network or on any remote servers.
- Distributed data—Learning data can be stored and shared across one or more systems.
- Usage and performance data—Data about learning resources that include not just quantitative metrics but also pedagogic context, skills, and performance.
- Team-based scenarios—Data associated with users can now be aggregated and associated with a team or group of users.
- Instructor and facilitator scenarios—Instructors or facilitators may observe and send or receive feedback or annotations to users during a learning or performance activity using real-time data collection displayed in an interface or dashboard.

The Shift to Adaptive Learning

A natural extension of big data and analytics is driving us toward what is referred to as adaptive learning, where technology mediates learning based on the unique needs of the individual learner. The primary driver behind adaptive learning was behavioral psychologist B.F. Skinner (Teasley n.d.). In the 1950s, Skinner created a "teaching machine" that focused on presenting new concepts to learners instead of reinforcing rote memorization. Today, the combination of data and technology enables connected devices, such as smartphones, tablets, and computers, to immediately adapt to the learner's input and context and present content in real-time based on that input. Alfred Remmits, a pioneer in the adaptive learning field, says, "Adaptive learning is the realization of machine learning and artificial intelligence. The concept is based on how large data sets are analyzed to provide detailed, specific, and relevant content experiences for the learner." An adaptive learning solution does the following:

- It continuously adapts and presents information based on learner input; for example, if the learner is answering questions correctly and is demonstrating mastery, positive feedback and reinforcement is displayed, and the content moves the learner to the next level. If the learner is having difficulty with the content, instead of simply prompting the learner to try the same questions again, the solution adapts and provides hints or different content.
- It provides a calculation of the learner's ability to grasp the content over time and, based on the thresholds set by the instructional designer, automatically presents reinforcing content to enable the learner to study.
- It offers continuous learning opportunities to appropriately challenge learners and confirm their capabilities.

Adaptive learning is the area in which big data plays an instrumental role in fashioning a learning strategy that helps individual learners to perform optimally. You now have a "capabilities heat map" of the existing workforce, enabling you to target specific areas of vulnerability, as well as recognize areas of high performance. This affects the entire life cycle of employee development, beginning with talent acquisition all the way through deep career progression. The core philosophy of adaptive learning is to leverage the trifecta of big data, machine learning, and artificial intelligence to provide more meaningful, relevant content for the learner. By bringing these three technologies together, you create a link between actual performance and desired performance, allowing you to surgically intervene. Gathering data is getting easier over time; interpreting the data is the challenge.

Ask the Expert: Alfred Remmits on Big Data and Learning Effectiveness

Alfred Remmits is the CEO of Xprtise, a workplace learning consultancy headquartered in Chicago, and also the European partner of Apply Synergies, a learning consulting firm. Alfred has recently been leveraging technologies such as big data to help companies deliver more effective learning.

How does big data affect the modern learning organization?

Big data will for the first time in history move the learning effectiveness needle from an activity-based measurement model (hours of training, certifications, and so forth), to an impact-based model. Over time, measurement will move to true ROI calculation based on business impact. We will be able to relate to the business KPIs and see the impact a broad scale of learning solutions will have on employee performance. Big data will also impact the learning organization, as it will require a data analysis skill set, and it will highlight the importance of the learning organization to the overall business. Once the impact of the learning interventions is aligned to the true KPIs and their value is calculated, the sky is the limit for the real business value the learning organization brings. Research has already shown that companies with a focus on measurement and data analytics perform significantly better.

How do learning organizations leverage data to become more aligned to business needs?

The real impact of this is noticeable when the learning organization transitions from a learning focus to a performance focus. When Bob Mosher and Conrad Gottfredson talk about the importance of focusing the learning solution on the moment of need, it means ensuring the workforce can be effective in their day-to-day performance. That is also where learning collects data, at the moment of need, showing the true impact of the learning design. So start with designing based on the problem to solve, the KPIs that need to be achieved, and base the solution on those two items. Move from offering a solution that solves problems and may achieve certain KPIs, to addressing the true business problem, defining the KPIs, and building the right solution. It sounds like a small shift, but by doing this right, the learning organization will begin to add demonstrable value to the business. The core component in this is aligning to the KPIs to achieve, and which will be truly measured based on the opportunity to collect the data.

How can we best choose to gather and use data extracted from learning experiences to provide demonstrable value to the business?

I believe that we should change the wording; let's not extract data from the learning experience, and instead, let's define the KPIs in the performance that need to change. We are going to measure what the change in performance KPIs will be based on the overall impact that the learning solution, not the learning experience, will have on the performance of the worker and the outcome of his or her work. Define the KPIs that need to be achieved as a starting point and then measure against them when deploying the complete solution.

How can learning teams leverage data and analytics to improve their overall design and development process?

The defined KPIs are the starting point, and big data will provide continuous (instead of a one-off) insight into the progress made with regard to achieving the required impact on the KPIs. The day-to-day business is becoming Agile in nature, and many organizations are starting to work in teams and on short-term projects. The learning organization is absolutely not aligned nor ready for this change and needs to adapt to the Agile development and deployment approach. I recently met with the learning organization of a large bank in the Netherlands and they informed me that their internal customer had moved to Agile and were asking for solutions to be delivered in weeks, and even days, whereas the learning organization could only deliver solutions within months based on the ADDIE model they worked under. This learning organization was completely out of sync with the business they support.

Is there a new skill set needed for learning teams to help them become a "data-driven" organization?

Yes, as I mentioned, the learning organization needs to adopt new skills in the data analytics space and start to use the business language by aligning the learning investment to the overall business KPIs. This way the learning organization will become a more strategic partner. Over time, they will become one of the key instruments to align resources based on changes in the market, competitors, and so forth.

Case Study: Using Data to Drive Employee Engagement

A work team's performance potential begins with their level of engagement. A highly engaged team is committed, energetic, and attached to the strategic imperatives. A disengaged team faces retention and performance challenges. Implementing consistent mechanisms to motivate employees is a key component in driving engagement. These are three of the most common challenges associated with disengaged employees (Glint 2015):

1. **Enable connectivity at all levels.** As the company or team grows, it's important to ensure the culture stays intact and all employees feel a sense of connectedness to the strategic goals and sense of purpose, as well as to each other.
2. **Be responsive during rapid growth or change.** For teams undergoing rapid growth or extreme change, it's imperative to foster cohesiveness. Teams can become fragmented, especially if there are frequent managerial changes. Even in the face of constant change, a sense of stability and continuity is critical.
3. **Be agile.** Fostering a work model that celebrates success while encouraging risk taking may be difficult in some organizations, but creativity and innovation are necessary. Empowering teams to take risks will drive employee commitment. Too much rigidity and unwillingness to consider different perspectives and viewpoints may cause stagnation.

Learning organizations have a unique role in driving engagement. Employees often rate training as the top benefit a company provides. Many learning organizations are focused on leveraging data to show evidence that their learning initiatives are driving higher levels of employee engagement. This case study highlights how a company with many employees began to look at training data through the lens of engagement and adjusted their evaluation methods to bring more insight into their ability to affect employee engagement.

It all began with a question from the CFO, who asked, "How do you know the training you create actually works?" That question motivated the learning team to dig into the collected data and seek evidence of learning effectiveness. Building on a solid foundation of data analytics, the learning team began to look at how and if the data could provide insight into employee engagement.

The learning technologist stated, "We analyzed enterprise-wide employee engagement data versus Kirkpatrick Level 1 data, which was obtained from all our formal learning offerings, to find out if the Level 1 data could provide insight into employee engagement."

The company surveys its employees once a year on their engagement level. Although employee engagement is a key performance indicator for the company, the survey is too infrequent to provide truly actionable data. Because the learning team can gather data continuously through their LMS, they felt more meaningful engagement data might potentially be uncovered. Next, the learning team began to work with other lines of business to collectively analyze the data.

The statistical analysis showed that Level 1 data strongly correlated to engagement. The team members selected courses with the lowest scores that reached the most employees. They temporarily implemented an open comments field to gather more qualitative data. They then reviewed the course content and used the qualitative data to determine how to improve the content. Additionally, this exercise led them to other data that uncovered relationships. They recently included two Level 1 (Reaction) surveys with their training program. The first survey asked questions about confidence and empowerment as it related to the training program. The second survey quantified engagement by inquiring on how the training program helped employees accomplish their job. The evaluation results showed a tight correlation between the two surveys, which helped team members use Level 1 metrics as leading indicators on the engagement survey. They expanded the Level 1 survey to ask Level 2 (Learning) and Level 3 (Behavior) questions, and discovered even greater opportunities to improve their training and employee performance.

The team plans to regularly seek relationships and evaluate established relationships at different times. Each month, members will investigate some of these data points to uncover areas of opportunity that they think may positively affect the learning experience. Every quarter, they further evaluate the two surveys to monitor the strength of the correlations and determine if they can continue to use this approach to affect engagement scores.

Operationalizing data interactions led the team to the next phase of data analysis, which focused on how to improve the learning design and development work flow.

The result of the data analysis led to the creation of a learning dashboard tied to the KPIs (including employee engagement), which is now regularly presented to the executive leadership team. They can now see a direct correlation between learning and employee engagement.

Ask the Expert: Ricardo Mejia on Learning Analytics

Ricardo Mejia, a learning technologist from the Home Depot, drives the learning effectiveness strategy for the company's learning organization. He leverages learning analytics to provide actionable information on the effectiveness of the learning the organization delivers to ensure that it's aligned to the business imperatives.

How do analytics improve the learning team's effectiveness?

Analytics turns on the lights for the learning organization. Without data and analytics, the organization is operating in the dark. Data-driven decision making is critical to forming an effective learning program. An analytics program is an essential component to a decision-making framework.

What's the best way to get your team started on a data-driven strategy?

The first step is to simply gain access to the data. Learn your data model and understand what's there. Next, begin to answer fundamental questions about the effectiveness of the learning you deliver. Then you can think about how to leverage technology to maximize the utility of the analytics. Invest time to determine how best to collect data and process it in a manner that yields high value with minimal effort. Let the machines do what they are good at (repetition and data crunching), while letting humans do what they are good at (creativity and organization). That dynamic will enable the team to amplify their productivity and better support the learning programs.

Should data inform every aspect of the learning organization's operations?

Data provides evidence of what occurs, informing decisions you then make. This leads to improvements in the learning you deliver. Data allows you to take chances where there may be little evidence, but where innovative approaches can yield substantial reward. This approach needs to be a key component in your learning measurement strategy, which can enrich the learning experience in a way that leverages operational efficiencies.

Key Takeaways

Applying a data analytics strategy for learning requires several steps to ensure success. First, focus on how to most effectively gather the data. You may need to work with your IT partners to establish a strategy on collecting learner data and determining the best business intelligence software for analysis. Second, identify the business challenges you need to address. Don't gather large data sets just because you can. Use the performance gaps or business problems you're trying to solve to lead you to data that matter. Third, determine how you will report the data. Dashboards and interactive tools offer different views to help visualize the issues, challenges, and potential next steps. Seeking a common data visualization strategy is an important component in effectively delivering the results of the data across the lines of business you support. You want to provide evidence to the business as to whether the learning solution resulted in value. Finally, it's critical to establish a data framework for ongoing reporting to show the value learning adds over time. A repeatable, consistent framework will improve insight and go a long way in helping the learning organization become a more valued partner with the business. Your data framework will be a living ecosystem, delivering crucial information for the business to act upon.

Data are everywhere. Your company is now one huge database, demanding that you become more data driven as a support organization. However, a data-driven learning organization is about more than the ability to extract data.

You must interpret the data and act on it. How do you make the information usable, browsable, and meaningful? In the same way you are transforming your learning designs away from static, one-time events, use your data analytics strategy to continually connect you to the business, its priorities, and its ever-changing needs.

5

The Internet of Everything

"Our expectation is that the Internet of Things will be much more than a screen and a keyboard."
—Vint Cerf, Chief Internet Evangelist, Google

SUZANNE LENTZ RUNS A SUCCESSFUL BAKERY, SPECIALIZING in wedding cakes. Her business has been growing, and she's feeling pressure to keep up with the orders coming in. In addition to hiring more employees, Suzanne is beginning to integrate smart appliances into her kitchen to increase productivity.

Suzanne started by installing a voice-activated digital assistant to give her a virtual set of hands in the kitchen. This new Internet-enabled device can start the dishwasher, preheat the oven, set timers, and even control a playlist of Suzanne's favorite songs. Assigning tasks such as these frees Suzanne and her staff to concentrate more on their core product, without spending time hunting for timers or setting dials. Suzanne's digital assistant even maps the fastest delivery routes and alerts her to the weather, which is vital when delivering wedding cakes. Restocking ingredients is also quick; all that's required is to ask the digital assistant and inventory is delivered in days. Reordering ingredients is also easier; all that's required is to ask her digital assistant and inventory is delivered in days.

With her smart device located in the kitchen, Suzanne is able to track tasks through the app even when she's not physically present. A smart kitchen is a good choice for Suzanne's bottom line because it helps her to more efficiently manage the rote tasks that used to take so much of her time. Smart objects are beginning to integrate into business in many ways, helping build connections that drive higher levels of capability, efficiency, and accuracy.

The Shift to Everything Connected

For more than 60 years, computers have assisted us with a variety of tasks. At first, computers were solitary machines used in large industries, the U.S. military, and in highly specialized fields for research. Since the 1970s, efforts to connect computers to one another became necessary to extract their true value. Vint Cerf, considered a father of the Internet, compared nonconnected computers to how business communicated in the early 20th century. Multiple phones in one office would each have different and often proprietary connections, making it difficult to communicate with a colleague in the same office (Greengard 2015). Fortunately, early Internet pioneers constructed a universal system to link computers instead of relying on closed systems, leading to the World Wide Web we know today.

Because the web's foundation is built on open standards, every object connects with a unique Internet Protocol (IP) address. Anything with an IP address can connect to another linked device. Unlike some other industries, the rapid adoption of open standards is what enabled the Internet to grow quickly and become widely used in such a short time. The founders of the Internet quickly agreed on a shared vision that, in less than 40 years, has set the foundation for ubiquitous connectivity. If you think about how long it takes for an invention (such as movable type, the printing press, transmission of sound, and broadcasting moving pictures through televisions) to transform our lives, it is remarkable that Internet connectivity occurred so rapidly.

The move to ubiquitous connectivity began in the late 1990s when Kevin Ashton, an assistant brand manager at Procter & Gamble, began working with radio frequency identification (RFID) chips to help manage supply chain operations. Ashton realized that a common protocol that would identify every component in the supply chain process could potentially increase efficiency and accuracy across the entire operation. By blending human input with machine input, and leveraging the strengths of humans and machines working together, Ashton believed the inefficiencies and errors of human-only processes could be greatly decreased or even eliminated. Ashton coined the term describing this process as the *Internet of Things* (IoT)—a way to bridge the gap between the physical and digital worlds.

The ability to connect objects extends far beyond our electronic devices. The idea of connecting things has quickly moved to the idea of connecting everything, including automobiles, tires, appliances, thermostats, lighting systems, parking meters, door locks, medical devices, farm equipment, clothing and fabric, tools, people, animals, trees, and more.

Consider a parking meter in almost any city today. Most meters now have sensors, capable of being connected, but do not by themselves respond to commands. However, access the parking meter using an app on your smartphone and it becomes an intelligent thing capable of more than just timing how long you've paid to park. The meter accesses the app to collect fees and generates reports on the back end. The smartphone has one IP address, capable of multiple functions used separately or together. The possibilities are infinite when multiple things are connected and addressable.

This connected universe, now referred to as the Internet of Everything (IoE), affords us near limitless capabilities to exploit the network effect to add features, services, and products to anything. The term *Internet of Everything*, created by Cisco Systems, is meant to denote the blending of both physical and digital things into a single entity—a connected experience—that spans the digital and physical space. As Samuel Greengard (2015, 34) states in his book *The Internet of Things*, "What makes the IoE so powerful is that it connects physical-first objects to each other as well as connecting them to digital-first objects, including computers and software applications." In essence, everything will have an IP address and be able to connect to a network, including you.

When you consider the power of everything connected, it's not a far reach to imagine the IoE will have a significantly larger impact on humanity than the Internet itself. The IoE evolution can be traced across three transformative occurrences—information, social, and physical—often referred to as graphs because of their interrelationships:

1. **Information graph**—the Internet created this graph, changing how we produce, access, share, and generate knowledge.
2. **Social graph**—ubiquitous connectivity led to the rise of social media, creating this graph, and changing how we establish and foster relationships with others.
3. **Physical graph**—the ability to connect everything creates this graph, changing how we interact with objects and environments.

This evolution brought a massive proliferation of connected, adaptive, and self-regulating things that will be able to learn and understand their context and adjust accordingly. Context is the key component in the IoE becoming more than just a bunch of discrete things and instead becoming a necessary part of our daily lives.

For example, one of the most challenging hurdles medical professionals face is their patients' inability or unwillingness to take prescribed medicine accurately and consistently. It's estimated that 75 percent of patients fail to take their medicine as prescribed, resulting in increased doctor visits and higher overall healthcare costs (Phrma 2011). Missed dosages of necessary medicine have massive financial implications to the healthcare profession, as well as an enormous cost on human health. With connected pills, the medicine itself can remind the patient when to take it, and sensors on the pill bottles can alert the pharmacy when to refill. The impact on the healthcare system of prescribed medicines taken more consistently is enormous. The savings alone to employer-based wellness programs, as well as reduced employee absences due to sick time off, is significant.

Additionally, you can start thinking of your house as a system instead of a collection of rooms. As you connect more smart things throughout your house, you monitor and regulate their connectivity; you want to make sure you don't configure a neighbor's device into your system, and vice versa. Almost all modern appliances now have more than just knobs and buttons; they have software and the ability to be connected. A connected washing machine can get information from outside your home, learn how to configure itself, and have a conversation with you about your laundry. Think of this as "laundry as a service." If you accidentally spill wine on your shirt, your washing machine can search how to remove the stain and adjust itself accordingly to make sure it removes the stain in the washing process (loads the right soap and sets the correct cycle). It's also able to track usage of soap and automatically place orders when inventory is low.

Both examples demonstrate how the IoE expands the way that we use software-enabled devices. However, it's important to consider the downsides of the IoE:

- Currently, device makers focus on just their device and might not consider other devices you have connected. In your house, you will eventually have multiple connected things, and you will want them to operate as an ensemble instead of discrete things unable or unwilling to work together. Almost half of the top global companies are platform companies, meaning that they have an array of products and services that form an ecosystem in which they operate. The platform companies are creating devices that interoperate only within their closed ecosystem, leading to fragmented systems. Without an open system driven by standards that enable scalability and interoperability, the IoE will devolve into a chain of closed systems with limited functionality.
- Respecting and clearly communicating individuals' rights to privacy and security as their devices converge is also important. Anything connected to the Internet is hackable. You must understand the implications for both business and personal use when more and more nonscreened devices begin connecting.

- Another challenge with having so many connected things is finding a way to power them without having to change batteries all the time or charge constantly. Advances in connectivity must be mirrored by progress in battery technology and creative ways to power things.

The Internet was fundamentally designed as a place for humans to talk to humans intermediated by machines. The IoE brings the machines more into the mix so that they're a part of the conversation, instead of just enabling the conversation.

Fundamentals of the IoE

The many benefits of the IoE for learning organizations begin with the simple fact that when everything is connected, these things can work together to help increase performance. Connected sensors and actuators interspersed in the work environment can assist both humans and machines to perform tasks, increasing productivity. The IoE is fundamentally driven by people, process, data, and things.

People

The transformative component of the IoE is the ability for people to always be participants in interactions with physical, but smart things. We are quickly adopting smartwatches, we will soon be wearing data-enabled apparel, we interact with connected medical devices, and we even ingest connected medicine to aid in our well-being. Although people will always play a part in the IoE, once artificial intelligence and machine learning technologies advance, connected things will operate more and more independently from human supervision and assume roles and responsibilities of their own.

Process

Everything that drives the IoE is tied together by process. Realizing innovative ideas and providing new capabilities for us to integrate into our lives is a fundamental reason for the rapid adoption of the IoE. This requires not only inventors but also developers to build software applications. For example, the Nest Internet-enabled thermostat is a smart thermostat designed to learn your habits and automatically adjust the temperature in your home. Developers build on its fundamental service to enable us to extract maximum value from the concept of a connected thermostat.

Data

The IoE changes not only how we collect data, but also what we can do with it. We created the Internet, first and foremost, as a way to share information. We love gathering and sharing data, and it will continue to be at the forefront of any new technology, platform, and paradigm we invent. All methods of communication (voice, text, and video) are just data, and historically we have generated, stored, and then shared data from one place to another, or transmitted it from one point to multiple points (broadcast radio or television). But now it's possible to transmit data to multiple streams simultaneously using the cloud. Swarms of data can now enable things to act autonomously, creating data colonies that can communicate to one another without a central network. Imagine an ecosystem of things monitoring almost any type of pattern and triggering action based on analysis of those patterns. Distributed, autonomous IoE ecosystems will monitor, regulate, and act upon conditions driven by a foundational component: data.

Things

Our things are the large and small objects, the devices, both simple and complex, that we continuously interact with and feed data into throughout our lives. We are forging an entirely new relationship with our things, and will eventually expect the majority of them to be smart, connected, and assistive.

Technology Fundamentals of the IoE

In its simplest form, the IoE is a platform that enables connectivity between things. A typical IoE platform architecture consists of the following components (Scully 2016):

- **Connectivity**–combine different protocols and data formats to enable interaction with all recognized things.
- **Things management**–ensure that the connected things are working properly.
- **Databases**–store data created by connected things.
- **Processing and action management**–analyze data from connected things and execute actions based on specific rules.
- **Analytics**–perform analysis from the data captured.
- **Visualization**–display patterns and trends from the data in a way that humans can interpret.

Robert Metcalfe, the co-inventor of Ethernet and founder of 3Com, stated in the 1980s, "The value of a network is proportional to the square of the number of connected users" (Scully 2016). The typical IoE architecture combines cloud technology and big data analytics along with various sensors from connected things to provide a continuous stream of real-time information. The data can span a wide variety of types and can trigger action from any of the connected things. The ensuing feedback loop can provide a host of services, capabilities, and interactions that engage the user and enable the experience to provide an individualized outcome. In this model, the IoE platform becomes an essential component in the experience, a means of gathering and making sense of the data, and using those data to optimize the experience.

There is immense potential for smart things to provide timely and valuable information through connectivity to help us make better decisions, mitigate risks, and add quality to our lives. Let's look at specific classes of connected things and their role in the IoE.

The Things in the IoE

Connected things are beginning to become an irreplaceable part of our lives in general, but they are also driving business in new ways. In certain industries, the IoE is leveraged to provide individualized customer experiences by offering relevant information when the customer needs it to help make purchasing decisions. Most of us have probably been in traditionally unstaffed spaces (airports, shopping mall walkways), where we are moving from one place to another, and noticed smart vending machines that provide specialty items.

In an airport, a vending machine may offer a choice between battery chargers for smartphones, headphones for listening to music on the plane, or extra cables for syncing devices. These vending machines are connected and communicate to the merchant about the data they gather from customers based on their needs and product selections. Merchants can communicate to the machines to dynamically adjust pricing based on weather conditions or other events that may attract interest around specific products. The machines also communicate inventory levels to vendors so they can be restocked accordingly.

Additionally, these machines can communicate to a passerby. If you're a repeat customer and a member of the company's loyalty program, the machine can send you text messages when you're nearby to prompt you about new products in which you're sure to be interested. The notification can even mention you by name to personalize the message. The machines constantly send data to merchants and vendors to keep them informed about regional, seasonal, and other microtrends in sales, so the machines can be stocked to meet specific demands. The choices in the Phoenix airport may be different from those in the Milwaukee airport to satisfy the types of customers walking by. The machine can further customize the experience by offering specific options to customers who have the company app installed on their smartphone. For example, if a desired product the customer is interested in is not available at this airport machine, the machine can inquire for one at the airport where the customer is flying to and confirm it can be picked up there, or if it's not available, have it shipped to the customer's destination to be picked up upon arrival.

This example represents the depth and breadth of connected things working harmoniously in a commerce transaction to drive business and customer satisfaction. An emerging trend with the IoE also revolves around the business, the customer, and the employee using the IoE in daily business transactions. As the IoE evolves, it's beginning to become infused in every nook and cranny of our work lives.

Wearables

Wearable technologies extend the reach and power of how people connect, communicate, and share details about their daily lives. Wearables respond in highly personalized ways, adapting their form and functionality to match a unique set of ever-changing needs. In business, early experiments in wearability have focused on the needs of highly mobile workers (such as couriers, logistics personnel, law enforcement officers, retail associates, and first responders), those whose jobs rely on timely access to situational information, yet require them to keep their hands free for other tasks (Greenfield 2016). The primary driver behind wearables in the workplace is to bring people into the data stream that occurs among connected things to drive productivity, efficiency, and improved information sharing. Wearables help make the worker more visible in the work stream itself, enabling the business to uncover areas of inefficiency and help streamline work flow.

Wearables typically fall into five categories: headwear, neckwear, armwear, bodywear, and footwear. Typical types of wearables used in business include smartglasses, wristbands, smartwatches, and badges. The armwear market has had the most promising use cases in business to date. For example, supply chain workers can wear armbands that track goods being transported through the distribution center, eliminating the time it takes to manually estimate and communicate loading times to managers. Instead, the data from the armbands are fed to software in the manager's office to provide real-time updating of the flow of goods through the distribution center. Similarly, warehouse workers wear GPS tags and use handheld scanners that tell them the most efficient route to take to collect an item for delivery to get it loaded as quickly as possible.

There are drawbacks to wearables as well. Anything connected to the Internet is hackable. Without rigorous encryption to safeguard data, breaches are possible. Too often, we focus on computers, tablets, and smartphones as the primary hackable devices; however, as everything becomes connected, most hackable things will be nonscreened devices. Wearables may start to be viewed as creating an oppressive or over-measured work environment in the eyes of workers. Employees may feel as though they're under the microscope, and may worry about personal measurements that can be captured, including heart rates, activity levels, location, and so forth. Companies need to clearly expand their privacy and data collection policies to include the IoE.

Sensors

Some technologists have said that the IoE is just the Internet of Sensors. The IoE requires sensors, such as RFID chips, to connect things to one another. In essence, sensors are what make things smart. They identify things, locate them, and determine environmental conditions (O'Donnell 2015). In the previous wearables examples, sensors are central to keeping track of the items in the warehouses. Sensors also live inside products to make them smart. Embedded sensors expose data, such as temperature, humidity, vibrations, motion, pressure, and so on, to enable data analysis of the situation and respond accordingly. Sensors are used not only for manufacturing and supply chain operations, but also to help customers find products and services. A customer wandering through a large retail store can quickly locate a product using an app on a smartphone if the retailer has enabled location technology. Sensing technology is the primary fuel behind the IoE. Typical sensor types include accelerometers, magnetometers, gyroscopes, acoustic sensors, pressure sensors, humidity sensors, temperature sensors, proximity sensors, image sensors, and light sensors. You can imagine the possibilities of all these sensors communicating to one another to provide information and analysis of real-time conditions. Sensors combined with other connected things generate massive amounts of data. In fact, the amount of data humans gather will never decrease; ever more data will be collected as everything becomes connected. The IoE exponentially increases the number of data sources along with the volume and variety of data to ensure our future is filled with data.

As we move to creating smarter things, sensors will continue to play a central role. Because of sensors, eventually your smartphone will have more advanced hearing and feeling capabilities than it does now, gaining a sense of smell and taste, and becoming even more context aware.

Ask the Expert: Oliver Kern on the IoE in Learning

Oliver Kern has been working with learning technology for more than 20 years, primarily focusing on how to apply new technology pragmatically to design and deliver effective learning. He designs and develops experiential learning opportunities, leveraging the IoE for government agencies and public companies.

How do you think the IoE transforms training organizations?

With the growing prevalence of sensors, bots, beacons, AI, cameras, and microphones, it's becoming easier to leverage technology to design and control experiential learning environments. For example, technology such as VR, AR, and AI are being integrated into corporate learning environments via virtual instructor-led training and message-based interactions with intelligent bots. It will take a while before these technologies become mainstream for the learning organization, but they'll become a key part of the learning landscape.

How important is it for learning leaders to focus on the IoE?

I think the IoE will take time to become a corporate learning staple, mainly because of its cost and complexity. It requires significant resources and capabilities to leverage it in a meaningful way. It will take time to identify its most productive applications.

How can the IoE affect skills and competencies?

If they're used to drive experiential learning, they can speed up learning and make it more sticky and fun. Skills could be learned and tested in a safe environment.

Have you seen other learning organizations implementing the IoE into their learning services?

I've seen some experimenting with beacons and sensors. Currently, it seems to be applied mostly in deeply experiential simulation training.

How does the IoE prove its value for learning?

With the IoE we have many means to measure effectiveness. We know that "smile" sheets delivered in classroom training reflect just how a participant feels about the training in the moment. This type of evaluation does not necessarily reflect the learning effectiveness or success of the training. With neutral sensors in simulated environments, on participants, in classrooms, and eventually in the workplace, we can potentially measure the success of learning and training more effectively.

How does the IoE help a person work with information more efficiently?

A person won't have to think about things that are present in their daily routines, such as driving, location logistics, supplies, communication, or meals. In other words, the IoE will make it easier for the individual learner to focus on the topic at hand. Focus is certainly one of the most crucial success factors for learning.

Learning as a Platform

The IoE presents a host of challenges and opportunities for learning organizations. First, there are the technological fundamentals to grasp; second, there are the things that make up the IoE to consider; and finally, understanding how to design learning experiences across multiple devices, including nonscreened devices, is a critical skill shift for learning teams. Once you have a good understanding of these fundamentals, you can determine if and how you leverage the IoE in your learning strategy. How you operationalize the IoE is a critical first step in supporting open, collaborative learning.

At its most fundamental level, the IoE helps training become more hyper-individualized, which leads to more meaningful and relevant learning experiences and directly helps the learner achieve desired outcomes. As John Hagel (2014) has stated, think of learning as moving from being a "knowledge store" to a "knowledge flow." Typically, when you design learning, you conduct an analysis of the workers and their environment, contrast the actual performance with the desired performance, identify the gaps in that performance you need to close, design the learning experience based on that analysis, and deliver it. You're spending money up front, pulling workers off the job to attend the training in person or virtually, and then hoping they'll perform in a way to justify the time.

What if you focused on connecting workers to people and systems that will help them in real time, driving performance improvement? You take a chance on getting the improvement up front and developing talent at a more rapid rate, instead of doing all the work ahead and hoping for an improvement on the back end. One critical component in ensuring an effective approach in this manner is to leverage the IoE to create learning as a platform—a distributed, interoperable ecosystem enabling people to build capability. The important element in approaching learning as a platform is how you harness data to provide an individualized approach, enabling learners to move at their own pace—faster, slower, or deeper, based on their own needs. Instead of providing a generic, one-size-fits-all learning experience, learning as a platform enables a true multidisciplinary approach to learning. This works by using data from connected things to monitor workplace performance, expose performance deficiencies, and identify where low, medium, and high performance occurs. Combine that with learners' data across their curriculum and you construct a more adaptive learning experience. Similar to platform companies like Amazon and Google, a well-designed learning platform encourages interaction among its users. By leveraging the peer-to-peer social learning model, you can foster deeper connections among the people on the platform and enable them to create deeper bonds with each other.

This is where your learning strategy completely changes. It's a move away from thinking of learning as a product and beginning to think of it as a platform. Your learning platform should include the following:

- affordances for the network effect (how many users, moderation, guidance mechanisms)
- their ecosystem of connected things, and how those things provide input into the overall learning experiences

- governance strategies (who may participate, how do you ensure ongoing value creation, and how do you resolve conflict).

The key is to not value the platform over the efficacy of the learning capability within the platform; instead, ensure the platform adds value where it's most needed.

The IoE and Learning Design

The IoE's technology focuses on the exchange of data between people and connected things, but learning organizations must look beyond the data and things to ensure that a cohesive learning experience results in the expected business outcomes. There is immense potential for smart things to work together to provide timely and valuable information to help workers in their moment of need; however, as with any learning technology, understanding the capability to provide effective learning should be the primary factor behind deciding to adopt it. Smart, connected things in the work environment can analyze data to predict worker behavior, anticipate problems and needs, and maximize the chances of a desirable result to help create a continuous learning environment that becomes better and more efficient.

Networked Learning

Instead of relying on transferring specific information to be retained, the learning platform facilitates richer learning that fosters critical thinking, deep collaboration, and social skills to help learners make decisions. With a large network available, learners can expand beyond their typical work silos and accelerate their learning. The platform enlarges the learning circle and creates environments conducive to improving performance more rapidly. This is where you stop investing heavily in up-front training programs that take people out of their work environment, and instead foster learning at work by leveraging connected things to support people in the motion of learning.

Taking the idea further of learning as a platform combined with IoE, you realize that learning how to do your job doesn't always happen within the four walls of where the job is performed. Learning doesn't stop when the worker goes home. When offering workers opportunities to continue learning beyond their work, you should construct mechanisms that enable capturing new capabilities back into the platform.

Interactive and Project-Based Learning

In this time of rapid change in the workplace, it's important to stress critical thinking skills with more interactive and project-based learning opportunities. Learning experiences should emphasize doing over passive presentation of information. With more learning experiences integrated into the work stream, you now have opportunities to create self-directed, hands-on learning facilitated by the smart things the worker is wearing, standing near, or holding. By leveraging the platform to provide customized coaching through the use of human or artificial intelligence, you help build expertise quicker by providing fine-grain content when its value can be best extracted. This enables learning continuity across devices, promotes collaboration between learners across the globe, and builds a global resource built on connectivity.

The Right Type of Learning at the Right Time

The IoE's biggest strength is in multichannel delivery: helping to determine the right type of learning at just the right time and present it to learners in a way in which they can easily interact. A true learning environment considers the context in which the learner is operating and provides information optimized for that context. By embedding information into the devices learners use to perform their jobs, you gain an advantage in being able to help learners at the time they're performing a task with which they're having trouble. This includes context-sensitive help built into the device, but also includes adding instruction to enrich the system, such as a direct video connection to help desks or vendor support. The best use of IoE for learning is to connect the thing and the person to ensure they both can perform optimally. This is where your learning strategy must broaden to embrace the learning as a platform philosophy. Most learning on the job occurs outside of formal training, so your strategy must now expand to include all the things with which workers interact while doing their jobs.

Beyond delivering learning through connected things, determining when to integrate other technology components is also key. The secret to getting people to engage with a learning experience is to reduce friction in the access of the experience—to make it as simple as possible. When thinking about multiple smart things, you should consider nonscreened things and potentially things that require voice input, such as chatbots. For some devices, the interface may be as minimal as a button or a few indicator lights. A work stream may also include multiple devices and services that may need to function together. Simplicity between things that are dedicated to one function may cause more complexity when considering how to design a learning experience around them. This is where artificial intelligence can play a prominent role in your learning design. Algorithms make sense of contextual data, anticipate user needs, and accept more natural forms of input, like voice commands. AI is not magical; it's just engineering. But to develop AI-based learning, you need the skills on your team to deeply understand the context in which AI can facilitate a learning experience. You also need to understand the constraints and possibilities of AI, as well as remain connected to the artifacts that make up the work environment.

The next big thing may not be a thing at all, but rather the connectivity of all things. The line between things is blurring. For the modern learning organization, realizing that learning is a part of work, not a separate event, is critical as you consider how technology is transforming the workplace. Only then will you reach the next phase in the evolution of the smart workplace.

Ask the Expert: David Blake on Learning as a Platform

David Blake is an education entrepreneur and the CEO of Degreed, a learning platform that leverages cloud-based technology to provide real-time learning resources. David describes Degreed as "the place that allows you to capture everything you're learning." David expounds on why learning is a platform, and how emerging technology affects today's workplace. He also discusses how workforce credentialing should be a critical component in any learning strategy.

What is your perspective on technology and how it affects workplace learning?

As we've engaged with our clients, two things stick out: the context of workforce automation and what technology impacts workers the most. For example, we're inclined to overestimate the impact of technology such as VR in the short term, while underestimating its impact in the long run. It will be quite some time before it influences anyone's daily life. Currently, VR is like massive online open courses (MOOCs) in that it's taking time for people to truly leverage the richness of the model. MOOCs are essentially a "sage on the stage" broadcast to large audiences.

Udacity is a good example of how to get your arms around a new format like MOOCs and iterate until it gets good. VR is in the early stages of determining what the learner experience consists of and the hurdles they must overcome to engage in the experience. Right now, people aren't comfortable with what's required for VR: It's still a stationary experience. VR will have its place in learning, but we are too quick to frame learning as always having to be an immersive experience. As VR evolves, we will be able to interact with each other, engage in simulations, and practice to achieve and demonstrate proficiency. To get there, we need nonstandard thinking and skills. Right now, as an industry, training is best equipped to administer standard skills at a scale. The pressure we face with more immediacy and poignancy, however, is how we support people who increasingly need nonstandard skills. We must shift to a model where as an industry we have the capabilities to help people by empowering them in the moment of need.

How can companies get more involved in building worker capabilities before workers join the company?

We need to take the premise that we will be better at empowering nonstandard skills. There's a difference between foundational learning and just-in-time learning. It's difficult to teach the nuances of JavaScript if the person doesn't have a web development foundation. We must think about both. The up-front foundation must be laid, and people need to know how to apply it in the moment of need to be truly empowered. The talent pipeline into the company is increasingly with people who don't have subject matter expertise and little to no foundation in the domain. We need to undertake two things as people come onboard: foundational knowledge and models of empowerment. As people join the company, transparency and relevance is critically important. We need to provide them the context for what they need to learn and why they should learn it. In the future, potential employees will learn about the company before joining. Companies will have their content, branded methodologies, and their learning opportunities available for the public to view. Employees will come into the company, already informed and prepared to work. I can take a strategy course to qualify myself as the person for them to hire. It's a way for me to qualify myself as an assistant store manager ahead of having the job. The threshold has come down. In the past, we thought of the corporate university as the place to learn about the job after we had the job.

What's your advice to learning leaders on integrating the five factors into their existing process?

For the most part, whether companies leverage technology is usually a function of their corporate IT policy. In our experience, it's been rare that learning leaders are afraid of technology; they're just usually confined by policy. However, this is the reality we're in; everyone must get their head around technology. It's been a decade since the iPhone was released, so along with BYOD, we now have "bring your own learning" (BYOL). If the company blocks access to YouTube on company devices, for example, people will bring their own device and access the services that provide the learning they need. People naturally access Google, YouTube, and other resources to get information they need. Trying to restrict access based on dubious IT policies only lasts so long because learning bleeds outside the organization. How do we bring value in this process, how do we bring people back to the organization? The command-and-control IT structure needs to be a command-and-control training function. Much of this is decided at the top and administered top down. Follow-up and accountability is a top-down function. Instead, learning organizations need to start thinking of a bottom-up learning strategy. How do we recognize what people are learning on their own outside of work? How do we empower people to get what they need when they need it, and give them credit?

Looking forward five years, where is technology and learning?

Over the next several years, we'll see the space continue to modularize and become more agile. We haven't historically been able to answer these questions: "What are our people learning?" "What are the skills of our competitors?" "What do people need to know?" Our workforces are all knowledge workers in a knowledge economy, but we don't know what they know. Technologies such as big data and the cloud help answer those questions, and five years from now, people will be able to tell you what they know, what their skills are, and identify paths to level up their skills. Big data and the cloud is the differentiator. Now we can capture data, leverage it, and establish career pathways, and then use the cloud to inform continuous content discovery.

If you were to design a corporate L&D learning strategy, what would be the first thing you would consider when it comes to technologies like the ones discussed in this book? What would you *not* do?

I think companies need to stop training generic skills. There's no reason to do this. Great content that teaches Microsoft Excel skills is not something a company should have to create and train. There are tons of resources freely available for this. People can go on their own to learn about Excel. Learning leaders should spend their time on the application of that skill in the context of the company. To get specific on developing content—if you are administering a program, ILT, or developing an online course, it should be specific to your company. If not, it shouldn't be a high priority. Timeframes most companies operate on are annual, semi-annual, or quarterly—not enough learning organizations think on a weekly or semi-weekly basis. Most of what your employees learned last year was not driven by the learning organization. They learned via podcasts, articles, videos, and their daily experiential learning. Learning organizations need to redefine the arc of their operation and get away from "do this in the spring and fall" and start talking about what their role is every day of every week.

Can you discuss how you think credentials affect the worker of today and in the future?

This is a topic that we're passionate about and deeply interested in. At Degreed, we divide the world between horizontal and vertical credentials. Horizontal credentials span across all employers, countries, and universities; for example, to provide a universal context for a credential. I will instantly know what the credential means if you tell me about it. For example, college credentials are universal. If I attain an MD or JD, people who don't know the specifics of my medical or legal expertise know what the actual credential means at a high level. Other credentials are vertical—they have value, but in a narrower context—such as a specific industry, role, or company. A Cisco network certified credential, an accounting credential, badging, a nanodegree achieved through a Udacity course, and so forth, are all examples of vertical credentials. Innovation will occur on both these axes. Verticals are constantly designing new badges, programs, and boot camps to deliver credentials. Degreed's mission is to unite credentials in a common format for all skills. Today, you ask a potential employee to tell you about their education, and you're going to view their last horizontal credential because that's all you have that has a universal expectation. That will change. Five years from now, they'll tell you about their level-six instructional designer credential, or their level-seven project manager credential.

What is the future of credentialing for corporate workers? Is everyone now a knowledge worker?

If credentials are currency, how widely accepted is that coin or token? Ultimately, there may be standard credentialing across industry sectors such as retail, manufacturing, and so forth, but credentials will need to cross several boundaries. For example, if I have a gardening credential, what does it mean? Is there a standard for this type of credential that employers will recognize? We will see standards, and we will attach them to things that matter.

Case Study: Intelligent Training Systems

One of the promises of the Internet of Everything is the ability to turn the workplace into an interactive, personalized learning environment. The Department of Defense created a fully immersive, game-based learning exercise designed to prepare soldiers to demonstrate resourcefulness under extreme pressure, such as a hostage-rescue situation. By using smartphones, video cameras, location and biometric sensors, beacons, and mobile apps, they could make the learning more realistic. The exercise was designed as an immersive simulation, conducted in the actual environment in which a rescue situation would occur. Trainees playing the roles of soldiers and hostage immersed themselves in roles to learn all aspects of a successful rescue attempt, as well as learning about collaboration, cooperation, and communication in high-stress situations.

Two days before the exercise, the instructional designers gathered data about the environment and the devices the trainees would wear, so that data could be gathered in real time during the exercise. This information was entered into the exercise software. The day of the exercise, the environment, the trainees, and their weapons were outfitted with sensors. Multiple sensors were attached to trainees to monitor their life systems, including pulse rate, heart rate, and skin temperature. Additionally, sensors were attached to their rifles, and video cameras were attached to their helmets. Sensors were also attached to objects in the environment, such as trees and buildings. Other sensors monitored humidity and wind. A proximity beacon was placed on the trainee who played the role of the hostage. All these connected things helped the trainees learn how to control their biometrics, develop their communication skills, and learn to cooperate to achieve the mission's objectives. The cameras attached to their helmets provided video feeds to the instructor's app, which provided detailed information about each trainee to an activity feed monitored by the instructor. Information from the sensors also provided the instructor with context analysis and trainee performance data. The data streams could be filtered to send instructor alerts when ranges were not met or to only highlight relevant information. The instructor could make instant field assessments and change the dynamics of the exercise at any moment. The opposing forces that trainees were to avoid were simulated in 3-D and rendered in the instructor's app, so he could see their movements against the trainees. The biometric sensors informed the trainees when unexpected conditions arose so they could react appropriately as they hunted for the hostage. When the hostage beacon was detected by the trainees, the opposing forces reacted. During the exercise, the instructor could increase the exercise difficulty to heighten the stress for the trainees, require them to collaborate and communicate more, or cause them to fail the mission. The instructor could add notes about each trainee and their responses into the system to use as assessment in the exercise debrief. The debriefs were immediate and detailed using the notes and activity data from the system.

This type of training exercise is a good example of an immersive experience that leverages the IoE to re-create a real-world learning exercise. In an experience such as this, there are four key design components to make it authentic:

1. **Reality layer**—the reality layer represents the environment in which learning will be applied. Learners won't remember the landscape or physical characteristics, but they will be able to remember the physical components.
2. **Ecosystem layer**—The ecosystem layer is where key decision points or nodes are mapped for an effective simulation. For

example, sensors can be placed in the ecosystem layer, and all the connected things (sensors, beacons, cameras, and so forth) create an ecosystem of nodes to provide data during the exercise.

3. **Game layer**—The game layer adds an artificial skin over the ecosystem layer. The game layer can be designed to reflect the reality layer for authenticity or to create a fantasy world. For example, if the reality layer is an open landscape, the game layer can place the learner in a medieval castle to trigger an emotional response and place the nodes closer to each other as game mechanics. Game mechanics ensure there's an emotionally heightened situation that will motivate learners to engage in the experience, as well as to provoke them to seek achievement.
4. **Data layer**—The data layer enables evaluation of all aspects of the performance and the performers. Quantitative data are collected from the sensors to understand context during assessment or to make the scenario dynamically adapt to the learners' performance. Qualitative data collected provide immediate assessment while learners are performing, enabling instructors to modify the experience, if necessary.

This type of intelligent training system enables richly interactive learning by leveraging technology where it's most applicable, assisting with the instructional intent. Immersive learning such as this example combines all five factors discussed in this book into a single learning solution: a cloud-based content management system, augmented reality, mobile apps and devices, multiple connected things, and data analytics.

Key Takeaways

Of all the factors discussed in this book, the IoE is the one with the most fuzziness for learning organizations, but it's also the one that offers more insertion points for learning to offer a comprehensive learning solution. As the workforce continues to wrestle with how to continually acquire the knowledge needed as work evolves, it's important to consider how you can move learning from a transactional, event-based model to a more embedded context-aware model that's available across channels. The overall move for your organization is to shift from being the holder and creator of a central knowledge store; instead, take your place as a leading contributor to the knowledge flow that's already cascading through the work environment.

When it comes to the IoE for learning organizations, consider these key points:

- The fundamental driver behind the IoE for learning is understanding people learn best when they're immersed in the environment in which they operate. The IoE affords the ability to deliver hyper-individualized, deeply authentic learning experiences to accelerate performance.
- The barrier to connectedness and connectivity has dropped, enabling more sophisticated cross-device capabilities.
- Rethink your learning strategy to position learning as a platform throughout your organization and leverage people, process, data, and things as you build a continuous culture of learning inside and outside the work environment.

Embracing ubiquitous connectivity and infusing it across your learning offerings will propel your learning organization forward so that you can be prepared to maximize the potential of worker capability across the spectrum of performance. In the end, it's about outcomes, and a forward-thinking, modern learning organization is anything but complacent when it comes to leveraging all the resources available to ensure a competent workforce can execute on the business imperatives. The IoE opens a new frontier of truly envisioning learning embedded into the very fabric of work.

6

Leveraging the 5 Factors

"We are drowning in information but starved for knowledge."
—John Naisbitt

IN JUST A SHORT TIME, WE HAVE moved from a facilitator-led, lecture-based learning model to a student-centered, collaborative learning model, incorporating technology almost faster than we can measure its effectiveness and implications on workplace performance. When businesses began offering in-house training more than 100 years ago, the typical structure was one of a top-down, controlled experience focused on skills-based competency for the industrial worker. Now, the advent of technologies such as the ones discussed in this book offer almost unlimited opportunities to meet learners where they are, at any time, with just the right information to help them become more competent. Alone or combined, each of the factors discussed in the previous chapters previews the learning organization of the future: offering cloud-based, multiplatform learning available at the tap of a finger; immersive, authentic experiences with real-time feedback and sharing capabilities; and data and real-time learning analytics to improve learning and provide evidence of workplace effectiveness.

According to John Seely Brown (2011) in his book *A New Culture of Learning: Cultivating the Imagination for a World of Constant Change*, the emerging learning paradox is based on three principles:

1. The old ways of learning are unable to keep up with our rapidly changing world.
2. New media forms are making peer-to-peer (P2P) learning easier and more natural.
3. P2P learning is amplified by emerging technologies that shape the collective nature of participation with those new media.

Putting the 5 Factors to Work

Over the next several years, the five factors will continue to bring dramatic change to how work is accomplished, challenging not only the modern worker, but also the modern learning organization. Emerging tools and technology, such as 3-D printing, digitized senses, gestural interaction, and drones as service agents, are establishing entire new platforms for our connected world. As technology evolves and becomes more sophisticated, complex relationships between workers and the technologies that support them in their work require you to provide a learning service that adds real value to the bottom line. Connecting the dots between the five factors and the worker is a significant challenge. Constructing a valued learning platform requires courageous leadership, an innovative spirit, and a concerted strategy to deliver maximum value at every point in the learning development life cycle. As you consider the five factors and their place in your learning strategy, remember these practical points:

- **Become IT-centric**. There will be obstacles in establishing a platform driven by both technology and learning effectiveness. Keep a sharp focus on these core principles: Generate and capture reliable, meaningful data and analytics that connect to business drivers; be creative in how you design and deliver learning; and know how the learners you support do their work.
- **Think of learning as a platform**. Delivering effective learning is now a multichannel operation, requiring a broad confluence of technologies and systems. To successfully leverage the five factors, you must first think of your learning service as a platform, one built on standards, connectable to the network, and able to receive input.

- **Build or buy your talent.** Your organization is itself building and sifting through massive amounts of information to construct meaningful learning experiences. Be sure to appropriately assess the skills across your team and provide the training and development your team needs to lead you forward. Your new learning strategy will not be operationalized overnight. You must develop the talent on your team to move ahead.
- **Break down the walls surrounding the outdated technology pervasive in your organization.** How can you support learners in the workplace if they carry more advanced technology in their pocket than you have in the company? Remember, there was a time when a typewriter on a worker's desk was the most advanced technology available.

Ask the Expert: Chad Udell

Chad Udell is a principal at Float Mobile Learning and has years of experience integrating technology into learning. Chad's recent focus has been discovering how to leverage many of the technologies discussed in this book to design and deliver optimal learning experiences for his clients. He has specific expertise in mobile, AR, and VR technologies.

How will connected devices, such as wearables and sensors, affect workplace productivity?

One of the more amazing possibilities that sensors and wearables bring to the workplace is the idea that employees could be assisted or provided new information, assistance, or performance support without requesting it or without even knowing they needed it.

How does a modern learning organization begin to integrate connected devices and smart objects into its learning strategy?

Proof-of-concept projects and pilot programs are always a great place to start. Begin one of these efforts with a single testable theory. Don't try to create an "enterprise-ready" solution right out of the gate. Ideate, test, and revise based on the changing technology and the feedback provided by users.

What does it mean for content development strategies when considering connected devices and smart objects from a training perspective?

You must distill the content to its simplest form. Remove the presentation layer or any associated behaviors embedded into the content. Only when the content is clean, a flexible metadata structure is in place, and a plan is ready for the future use of content that is separate from the user interfaces and user experiences that it may be used in, should you consider your content strategy off and running.

How does IoE help drive learning effectiveness?

Whether it can or not remains to be seen, as it is such a new area of exploration. It could theoretically drive learning effectiveness by enhancing the content relevancy and reducing the friction to access it. Both of these would undoubtedly be positive contributing factors in delivering more effective learning solutions.

Why should a learning leader care about technologies such as the IoE?

The mass deployment of workplace IoE is not a question of if, but rather of when. IoE will transform workplace productivity. If learning leaders are not equipped or versed in emerging enterprise technology trends, their teams and overall value to the organization as a whole are diminished. As the workplace changes, so should learning.

What does it mean for a modern learning organization to be data driven?

To me, to be data driven means that an organization makes changes to their offerings based on fact, rather than feeling. Certainly intuition is valuable, as it can often help create innovation and move things a little more rapidly than simply being reactive. The contrast to that is an organization steeped in data to the point that the reactionary stance taken by someone reviewing customary or noninsightful reports is replaced with a person or algorithm primed to proactively manage processes, people, and tools.

What is the best way for a learning leader to think about analytics from a learning strategy perspective?

A simple example is if a worker's behavior is something you aim to change, you will need a way to measure that behavior. Once a tool or rubric for measuring that behavior has been established, a correlating learning tool or content piece is going to be created. When this is done at scale, and the behaviors and learning content can be matched up in a reliable fashion, you have the metrics in place that can contribute to a meaningful set of data that can be used for learning analytics.

What change management is required for a learning organization to make the shift to being data driven?

The thought *this is the way we've always done it* is not only hyperbolic but also dangerous. Change cannot come without accepting new ways of doing things. New ways of doing things require planning and agreement around the value proposition these new things will bring. When you are intuition driven rather than data driven, you are, by definition, opposed to change.

We can now gather massive amounts of data about what our learners are doing. What's necessary to appropriately filter the data to get to meaningful actions we can take to get to real results?

You must begin with the goals in mind. It's very easy to capture everything. Resisting that urge and keeping your focus on the information that is most important is a necessary first step.

What role in a data-driven learning organization does today's LMS play?

The LMS at your organization may be able to provide a role in the transformation to being a data-driven organization, but it shouldn't be considered vital. If your LMS is really only capable of capturing traditional learning data, it may not be scalable or flexible enough to gather the large amounts of data needed to enter the realm of predictive analytics. There are many other platforms that facilitate large-scale data gathering and also assist in the delivery of meaningful analytics reporting tools that aren't even in the learning world. It may be worth a bit of time exploring alternatives.

What role should a learning leader play in consideration of new learning technologies?

Learning leaders have the opportunity to be the evangelist, the skeptic, the coach, and the experimenter all at once. It's important that they keep a watchful eye on technology and manage their interplay to match or push their organization's culture accordingly. Communication with the variety of stakeholders on this topic should be tailored to meet the audience's needs as well.

How should a modern learning organization view the five factors discussed in this book as part of their overall strategy?

A vital part of understanding how to integrate technology into the learning flow is to gain better understanding of the worker's context—the time, setting, and intent in which information may be needed to be accessed. This is why a human-centered design approach is paramount. Once you have the context fully grasped, you can begin to empathize with the learner in a much more real fashion. This will also help in crafting content, activities, and utilities that enable ambient and intelligent interventions. Maybe you can create and deliver something to workers that they didn't even know they wanted or needed.

Do you have any further thought on how the five factors affect workplace learning?

In our ever-changing workplace, many content delivery platforms and/or interaction points may not have a screen at all. In this case, plan for how workers access or are provided the information and how to offer the most value in a way that doesn't require them to sit in front of a screen.

Are You Ready for the Age of Immediacy?

Technology itself is meaningless if it doesn't make our lives better. Ultimately, it's humans who build, design, and use technology to get work done. We have experienced a fundamental shift in how quickly new technology is adopted, in how workers improve their skills and knowledge, and how we collectively view work itself. Learning is now driven first from the individual's needs, has become more self-directed, and must be relevant and meaningful. Learning in the Age of Immediacy is defined by its lack of limitations and lack of barriers to obtaining information and knowledge at the time of need. Traditionally, corporate learning organizations are not designed for quick and radical change. However, we have entered an era where the most important skill is to acquire new skills. In the rear-view mirror is the reflection of how you used to do training. The Age of Immediacy requires learning organizations to be less reactive and more nimble, flexible, proactive, and directly aligned to business drivers. More and more, the business is looking to the learning organization to help ensure that the workforce operates competitively and can execute on their ever-shifting imperatives. To ensure workforce readiness, innovative and creative solutions are necessary, more exacting results and deeper analysis is required, and the support and resources the modern worker needs must be delivered in a timely manner. As you transform your learning organization, look to the edges, around corners, and beyond the blind spots that inhibit innovation. Seek different experiences. Spend time with people outside your network and industry. Connect closely with your learners and customers to increase awareness of the changes that are critical to move your organization ahead.

It's exciting, but daunting. You are now faced with constructing the next-generation learning organization, reinventing how you design and deliver training by evaluating the technologies, tools, and methods that will move you forward. Because of the scale of this transformation, thriving in the Age of Immediacy represents an enormous opportunity to do more, learn more, be more productive and effective, and positively affect your customers, your business, and those who rely on you for the skills and knowledge they need to do their jobs. From now on, you must have an ongoing, real-time dialogue with your learners. We're all connected—there is no concept of offline anymore. Welcome to the Age of Immediacy.

References

Abowd, G. 2016. "Beyond Weiser: From Ubiquitous to Collective Computing." *Computer* 49(1): 17-23.

Belissent, J. 2016. "Improving the Reality of Government: AR and VR Use Cases." Forrester Blog, May 9. www.blogs.forrester.com/jennifer_belissent_phd/16-05-09-improving_the_reality_of_government_ar_and_vr_use_cases.

Bellini, H., W. Chen, M. Sugiyama, M. Shin, S. Alam, and D. Takayama. 2016. "Profiles in Innovation: Virtual & Augmented Reality." Goldman Sachs, January 13. www.goldmansachs.com/our-thinking/pages/technology-driving-innovation-folder/virtual-and-augmented-reality/report.pdf.

Bloom, S. 1984. *Taxonomy of Educational Objectives Book 1: Cognitive Domain*, 2nd edition. Boston: Addison-Wesley Publishing Company.

Bort, J. 2016. "Secret Passages and Skipped Meals: Oracle's CEO Gave Us a Rare Peek at What It Really Takes to Run a $37 Billion Company." *Business Insider*, September 26. www.businessinsider.com/oracle-ceo-mark-hurd-openworld-day-in-the-life-2016-9.

Brown, J., and D. Thomas. 2011. *A New Culture of Learning: Cultivating the Imagination for a World of Constant Change*.

Brown, T. 2008. "Definitions of Design Thinking." Design Thinking Blog, September 7. www.designthinking.ideo.com/?p=49.

Cerf, V. 2016. "Safety, Security and Privacy in the Internet of Things." Presentation at the Internet of Things Conference, February 18. Mountain View, CA: Internet Society. www.sfbayisoc.org/iot-conference.

Checkpoint Systems. n.d. "RFID Heritage." Checkpoint Systems. http://us.checkpointsystems.com/solutions/merchandise-availability-solutions/merchandise-visibility/rfid-heritage.

Conan Doyle, A. 2011. *Collected Works of Sir Arthur Conan Doyle*, 6th edition. Hastings, East Sussex: Delphi Classics.

Evans, B. 2016. "Mobile Is Eating the World." Benedict Evans Blog, December 9. www.ben-evans.com/benedictevans/2016/12/8/mobile-is-eating-the-world.

Farber, D. 2008. "Oracle's Ellison Nails Cloud Computing." CNet, September 26. www.cnet.com/news/oracles-ellison-nails-cloud-computing.

Gayomali, C. 2014. "Why Did Google Purchase Artificial Intelligence-Firm DeepMind?" *Fast Company*, January 27. www.fastcompany.com/3025518/tech-forecast/why-did-google-purchase-artificial-intelligence-firm-deepmind.

Glint. 2015. *The Chemistry of Employee Engagement*. Redwood City, CA: Glint.

Greenfield, A. 2006. *Everyware: The Dawning Age of Ubiquitous Computing*. New York: Pearson Education.

Greengard, S. 2015. *The Internet of Things*. Cambridge, MA: MIT Press.

Hagel, J. 2014. "Workplace Redesign: The Big Shift From Efficiency." Presentation at SXSW, March 7. Austin, TX: SXSW.

Harrington, L., and M. Heidkamp. 2013. "The Aging Workforce: Challenges for the Health Care Industry Workforce." The NTAR Leadership Center, March. www.dol.gov/odep/pdf/ntar-agingworkforcehealthcare.pdf.

Harris, S.D. 2009. "Q&A: Marc Benioff, CEO of Salesforce.com." *San Jose Mercury News*, October 23. Updated August 13, 2016. www.mercurynews.com/2009/10/23/2009-qa-marc-benioff-ceo-of-salesforce-com.

Hawkins, J., and D. Dubinsky. 2016. "Let's Reverse-Engineer the Brain." Audio podcast, interview by Kara Swisher. Reccode Decode, June 27. www.recode.net/2016/6/27/12037248/artificial-intelligence-machine-learning-numenta-jeff-hawkins-donna-dubinsky-podcast.

Helft, M. 2016. "One-on-One With Google CEO Sundar Pichai: AI, Hardware, Monetization and the Future of Search." *Forbes*, May 20. www.forbes.com/sites/miguelhelft/2016/05/20/one-on-one-with-sundar-pichai-on-the-future-of-google/#1e517636ce81.

Heskett, J., T. Jones, G. Loveman, E. Sasser Jr., and L. Schlesinger. 2008. "Putting the Service Chain Model to Work." *Harvard Business Review*, July. www.hbr.org/2008/07/putting-the-service-profit-chain-to-work.

Hodges, A. n.d. "The Alan Turing Internet Scrapbook." Alan Turing: The Enigma. www.turing.org.uk/scrapbook/test.html.

Hoffman, R. 2016. "Using Artificial Intelligence to Set Information Free." *MIT Sloan Management Review*, June 14. www.sloanreview.mit.edu/article/using-artificial-intelligence-to-humanize-management-and-set-information-free.

Investopedia. 2011a. "Cloud Computing." Investopedia, January 18. www.investopedia .com/terms/c/cloud-computing.asp.

——. 2011b. "Software as a Service - Saas." Investopedia, January 4. www.investopedia. com/terms/s/software-as-a-service-saas.asp#ixzz4IYJ9Ow7e.

Kelly, K. 2016. *The Inevitable: Understanding the 12 Technological Forces That Will Shape Our Future*. New York: Penguin Publishing Group.

Kirkpatrick, D., and J. Kirkpatrick. 2006. *Evaluating Training Programs: The Four Levels*, 3rd edition. San Francisco: Berrett-Koehler.

Kiron, D., P. Prentice, and R. Ferguson. 2014. "The Analytics Mandate." *MIT Sloan Management Review*, May 12. www.sloanreview.mit.edu/projects/analytics-mandate.

Kissane, E. 2011. *The Elements of Content Strategy*. A Book Apart. https://abookapart .com/products/the-elements-of-content-strategy.

Levine, S., T. Lillicrap, and M. Kalakrishnan. 2016. "How Robots Can Acquire New Skills From Their Shared Experience." Google Research Blog, October 3. https:// research.googleblog.com/2016/10/how-robots-can-acquire-new-skills-from.html.

Lovinger, R. 2007. "Content Strategy: The Philosophy of Data." Boxes and Arrows, March 27. www.boxesandarrows.com/content-strategy-the-philosophy-of-data.

Marr, B. 2016. "Will 'Analytics on the Edge' Be the Future of Big Data?" *Forbes*, August 23. www.forbes.com/sites/bernardmarr/2016/08/23/will-analytics-on-the-edge-be-the-future-of-big-data/#2b05a1bd2b09.

McKalin, V. 2014. "Augmented Reality vs. Virtual Reality: What Are the Differences and Similarities?" *Tech Times*, April 6. www.techtimes.com /articles/5078/20140406/augmented-reality-vs-virtual-reality-what-are-the-differences-and-similarities.htm.

Murphy, K. 2015. "Feeling Woozy? It May Be Cyber Sickness." *New York Times*, November 14. www.well.blogs.nytimes.com/2015/11/14/feeling-woozy-it-may-be-cyber-sickness/?_r=1.

Nash, J. 1997. "Wiring the Jet Set." *Wired* magazine, October 1. www.wired .com/1997/10/wiring.

O'Donnell, J. 2015. "How Smart Sensors Are Transforming the Internet of Things." Tech Target, July. http://internetofthingsagenda.techtarget.com/opinion /How-smart-sensors-are-transforming-the-Internet-of-Things.

Perlis, A. 1982. "Epigrams in Programming." Yale University. www.cs.yale.edu /homes/perlis-alan/quotes.html.

Phrma. 2011. *Improving Prescription Medicine Adherence Is Key to Better Health Care.* Washington, D.C.: Phrma. www.phrma.org/sites/default/files/pdf/PhRMA_ Improving%20Medication%20Adherence_Issue%20Brief.pdf.

Porter, M., and J.E. Heppelmann. 2015. "How Smart, Connected Products Are Transforming Companies." *Harvard Business Review*, October. www.hbr .org/2015/10/how-smart-connected-products-are-transforming-companies.

Rosenbush, S., and C. Boulton. 2013. "The Automation of IT Is Changing the Way Companies Innovate." *Wall Street Journal*, August 13. http://blogs.wsj.com /cio/2013/08/13/the-automation-of-it-is-changing-the-way-companies-innovate.

Scully, P. 2016. "5 Things to Know About the IoT Platform Ecosystem." *IOT Analytics*, January 26. www.iot-analytics.com/5-things-know-about-iot-platform.

Smith, A. 2016. "Public Predictions for the Future of Workforce Automation." Pew Research Center, March 10. www.pewinternet.org/2016/03/10/public-predictions-for-the-future-of-workforce-automation.

Sull, D. 2009. "Competing Through Organizational Agility." McKinsey & Company, December. www.mckinsey.com/business-functions/organization/our-insights /competing-through-organizational-agility.

Teasley, D. n.d. "Adaptive Learning: Definition, History & Methodology." Study. com. www.study.com/academy/lesson/adaptive-learning-definition-history-methodology.html.

Tobe, F. 2016. "AUTOMATICA 2016: Digitalization and Collaborative Robotics." *The Robot Report*, July 24. www.therobotreport.com/news/automatica-2016-digitization-collaboration-and-service-robotics.

Toffler, A. 1980. *The Third Wave*. New York: Morrow.

Vasic, M., and A. Billard. 2013. "Safety Issues in Human-Robot Interaction." In *IEEE International Conference on Robotics and Automation*, 197–204. Piscataway, NJ: IEEE.

Weiser, M. 1996. "Ubiquitous Computing." Ubiq.com, March 17. www.ubiq.com /hypertext/weiser/UbiHome.html.

Wikipedia. 2016. "Mainframe Computer." Wikipedia. https://en.wikipedia.org/wiki /Mainframe_computer.

Wood, R. 2015. "Robots Among Us." *MIT Technology Review*, November 3. www .events.technologyreview.com/emtech/15/video/watch/rob-wood-robots-among-us.

Yoshiaki, N. 2015. "In Japan, the Rise of Machines Solves Labor Shortage." *Bloomberg* News, September 13. www.bloomberg.com/news/articles/2015-09-13/in-japan-the-rise-of-the-machines-solves-labor-and-productivity.

About the Contributors

Gregory Abowd

Gregory Abowd is the J.Z. Liang Professor in the School of Interactive Computing at the Georgia Institute of Technology. He is a computer scientist best known for his work in ubiquitous computing, software engineering, and technologies for autism.

David Blake

David Blake is the co-founder and CEO of Degreed, an award-winning learning platform built for the way today's workers build skills and grow their careers. Prior to Degreed, he helped launch a competency-based, accredited university and was a founding team member at Zinch (acquired by Chegg). David was selected as a Top EdTech Entrepreneur by the Stanford d.School EdTech Lab, and has been published in *Harvard Technology Review*, *Business Insider*, *TechCrunch*, and the *Huffington Post*. He has spoken around the world on the topic of the future of learning, including at the ASU Education Innovation Summit, EdTech Europe, and TEDx.

Jason Haag

Jason Haag has more than 15 years of experience in education and training technology and distributed learning systems. He spent eight years supporting the U.S. Navy's e-learning program in both engineering and management roles before joining the Advanced Distributed Learning (ADL) Initiative in 2009. He is currently employed by the Tolliver Group and provides systems engineering and technical adviser support for the ADL, sponsored by the Office of the Under Secretary of Defense for Personnel and Readiness.

John Hagel

John Hagel is co-chairman for Deloitte LLP's Center for the Edge, with nearly 30 years of experience as a management consultant, author, speaker, and entrepreneur. He has served as senior vice president of strategy at Atari and is the founder of two Silicon Valley startups. He is co-author of the books *The Power of Pull: How Small Moves, Smartly Made, Can Set Big Things in Motion; Net Gain: Expanding Markets Through Virtual Communities; Net Worth: Shaping Markets When Customers Make the Rules; Out of the Box: Strategies for Achieving Profits Today and Growth Tomorrow Through Web Services;* and *The Only Sustainable Edge: Why Business Strategy Depends on Productive Friction and Dynamic Specialization.* John holds a BA from Wesleyan University, a BPhil from Oxford University, and a JD and an MBA from Harvard University.

Oliver Kern

Oliver Kern specializes in marketing, innovation, and communication. At Bayer, where he has been for 20 years, he established SkillCamp—a global marketing and sales learning framework—for the CropScience division. His focus areas are web strategy, innovation management, financial instruments, academy or learning programs, performance support, consulting, and coaching. He holds a degree in photographic engineering and an executive MBA with Kellogg and WHU. Oliver has a keen interest in curiosity (www.staycurious.tv) and deep experience in the creative process and creative problem solving; he is a certified consultant in FourSight and MBTI.

Tom King

Tom King is an independent learning strategist and consultant. He was instrumental in specifying and creating early implementations of ADL SCORM, AICC, IEEE LTSC, LETSI, and other learning standards. He has worked with Accenture, Macromedia, Adobe, Boeing, and the MASIE Center.

Seth Malcolm

Seth Malcolm is a senior operations program manager at Microsoft supporting IT, where he focuses on cloud technologies and enterprise networking services. Malcom is an expert on IT infrastructure technology and its impact on business.

Ricardo Mejia

Ricardo Mejia leads the learning technology team at the Home Depot. His charter includes managing the learning platforms, analytics, effectiveness measurements, and reporting for the enterprise's learning initiatives. Prior to the Home Depot, Ricardo was a business consultant for companies such as Hewlett-Packard, Procter & Gamble, CNA, and Sprint. He is a graduate of the Georgia Institute of Technology, with a degree in engineering.

Alfred Remmits

Alfred Remmits is a European entrepreneur and investor in learning technology, with a focus on workplace learning, performance support, adaptive learning, and artificial intelligence technologies. He founded several leading learning and performance companies in both Europe and the United States. He is currently the CEO of Xprtise, with headquarters in Chicago, and also the European partner of Apply Synergies.

Enzo Silva

Enzo Silva, learning strategist for SAP, leads the design and development of learning experiences mediated by emerging and trending technology and frameworks. He has also spearheaded a community aimed at the systematic adoption of innovations in talent development at SAP. His interests span social media, virtual worlds, gamification, games, language learning, acting, and helping others learn and apply creative ideas and tools to solving problems.

Chad Udell

Chad Udell is managing director at Float (gowithfloat.com). He works with both small and large learning organizations to explore and implement new technology into their learning services. Chad has written and published two books on mobile learning, *Learning Everywhere: How Mobile Learning Is Transforming Training* and *Mastering Mobile Learning: Tips and Techniques for Success.*

About the Author

BRANDON CARSON IS AN INNOVATIVE LEARNING STRATEGIST with extensive experience leading corporate global learning programs and teams in environments as varied as startups, tech companies, and retail. He is a popular speaker, delivering a wide variety of engaging presentations and workshops at industry events.

Brandon has been honored with several learning industry awards, including the ATD BEST Award, the eLearning Guild People's Choice Award, and three Brandon Hall Awards for Best Custom Learning Design. He has also served on the board of the North American Simulation and Gaming Association and ATD conference committees.

Brandon invites you to chat with him on Twitter (@brandonwcarson) and on LinkedIn (www.linkedin.com/in/brandoncarson), where he engages in frank, interactive discussions about learning and business transformation.

Index